1969

h's book ...

HARMONIC ANALYSIS AND THE THEORY OF PROBABILITY

HARMONIC ANALYSIS AND THE THEORY OF PROBABILITY

BY

SALOMON BOCHNER

UNIVERSITY OF CALIFORNIA PRESS
BERKELEY AND LOS ANGELES
1960

UNIVERSITY OF CALIFORNIA PRESS
BERKELEY AND LOS ANGELES
CALIFORNIA

CAMBRIDGE UNIVERSITY PRESS
LONDON, ENGLAND

Second Printing

This book was written with the partial support of the Office of Naval Research, United States Navy, under contract with the Statistical Laboratory, University of California, Berkeley, California.

Printed by lithography in the United States of America

PREFACE

This is a tract on some topics in Fourier analysis of finitely and infinitely many variables and on some topics in the theory of probability and the connection between the two is a very intimate one on the whole.

Although drafted in part earlier, more than half of the tract was actually written while the author was visiting, February–August 1953, the Statistical Laboratory at the University of California, Berkeley, of which Dr Jerzy Neyman is the Director, and a most delightful and profitable visit it was.

Special thanks are due to Dr Loève for listening patiently to expoundings of half-ready results, and to Mrs Julia Rubalcava, also of the Laboratory, for preparing the typed copy of the entire manuscript.

S. B.

CONTENTS

CHAPTER 1

APPROXIMATIONS

1.1. Approximation of functions at points

In ordinary Euclidean space E_k: $(\xi_1, \ldots, \xi_k)$,

$$-\infty < \xi_j < \infty, \quad j = 1, \ldots, k,$$

for any dimension $k \geqq 1$, the ordinary Lebesgue measure element $d\xi_1 \ldots d\xi_k$ will usually be denoted by dv_ξ. A function $f(\xi_1, \ldots, \xi_k)$ will also be written briefly as $f(\xi)$ or $f(\xi_j)$, and we will also put

$$|\xi| = (\xi_1^2 + \ldots + \xi_k^2)^{\frac{1}{2}}.$$

We take a family of functions

$$\{K_R(\xi_1, \ldots, \xi_k)\}, \tag{1.1.1}$$

also called 'kernels', subject to the following assumptions. The index R ranges over $0 < R < \infty$ and has continuous and occasionally only integer values. For each R, $K_R(\xi_j)$ is defined and Lebesgue integrable over E_k, so that the integrals

$$\int_{E_k} K_R(\xi_j)\, dv_\xi, \quad \int_{E_k} |K_R(\xi_j)|\, dv_\xi$$

exist, and we have

$$\int_{E_k} K_R(\xi_j)\, dv_\xi = 1 \tag{1.1.2}$$

for all R, and

$$\int_{E_k} |K_R(\xi_j)|\, dv_\xi \leqq K_0 \tag{1.1.3}$$

with K_0 independent of R; and, what is decisive, for each $\delta > 0$, no matter how small, we have

$$\lim_{R \to \infty} \int_{|\xi| \geqq \delta} |K_R(\xi_j)|\, dv_\xi = 0. \tag{1.1.4}$$

We note that for $K_R(\xi_j) \geqq 0$, (1.1.2) implies (1.1.3) with $K_0 = 1$.

Starting from an integrable function $K(\xi_1, \ldots, \xi_k)$ with

$$\int_{E_k} K(\xi_1, \ldots, \xi_k)\, dv_\xi = 1,$$

if we put

$$K_R(\xi_1, \ldots, \xi_k) = R^k K(R\xi_1, \ldots, R\xi_k), \tag{1.1.5}$$

then this is a family as just described, since by the change of variables $R\xi_j \to \xi_j$, $j=1,\dots,k$, we obtain

$$\int_{E_k} K_R(\xi_j)\,dv_\xi \equiv \int_{E_k} K(\xi_j)\,dv_\xi = 1,$$

$$\int_{E_k} |K_R(\xi_j)|\,dv_\xi = \int_{E_k} |K(\xi_j)|\,dv_\xi = K_0,$$

$$\int_{|\xi|\geqq\delta} |K_R(\xi)|\,dv_\xi = \int_{|\xi|\geqq R\delta} |K(\xi_j)|\,dv_\xi,$$

and for fixed $\delta>0$ the point set $\{|\xi|\geqq R\delta\}$ converges to the empty set as $R\to\infty$. Sometimes a statement will be intended only for such a special family of kernels, as will be indicated by the context.

For a measurable function $f(x)\equiv f(x_1,\dots,x_k)$ in E_k we introduce, if definable, the approximating functions

$$s_R(x) = \int_{E_k} f(x_1-\xi_1,\dots,x_k-\xi_k)\,K_R(\xi_1,\dots,\xi_k)\,dv_\xi; \qquad (1.1.6)$$

and since (1.1.2) implies

$$f(x) = \int_{E_k} f(x) K_R(\xi_1,\dots,\xi_k)\,dv_\xi,$$

we obtain $$s_R(x)-f(x) = \int_{E_k} (f(x-\xi)-f(x))\,K_R(\xi)\,dv_\xi,$$

and our first statement is as follows:

THEOREM 1.1.1. *If $f(x)$ is bounded in E_k*

$$|f(x)| \leqq M, \qquad (1.1.7)$$

then $$s_R(x)\to f(x),\ R\to\infty$$

at every point x at which $f(x)$ is continuous Also, if $f(x)$ is continuous in an open set A, then the convergence is uniform in every compact subset $\overline{A}_0$.

Proof. We have

$$|s_R(x)-f(x)| \leqq \int_{E_k} |f(x-\xi)-f(x)|\,.\,|K_R(\xi)|\,dv_\xi$$
$$= \int_{|\xi|<\delta} + \int_{|\xi|\geqq\delta} = I_1(R,x)+I_2(R,x).$$

Now, $$|I_1(R,x)| \leqq \sup_{|\xi|<\delta} |f(x-\xi)-f(x)|\,.\,K_0,$$

and by continuity of $f(x)$ at x this is small for δ sufficiently small. However, for δ fixed (small) we have

$$|I_2(R,\xi)| \leqq \int_{|\xi|>\delta} (|f(x-\xi)|+|f(x)|) \, . \, |K_R(\xi)| \, dv_\xi,$$

and by (1.1.7) this is $\leqq 2M \int_{|\xi| \geqq \delta} |K_R(\xi)| \, dv_\xi$, which is small for large R by explicit assumption (1.1.4), q.e.d.

The global requirement (1.1.7) was only needed for obtaining

$$\lim_{R \to \infty} \int_{|\xi| \geqq \delta} |f(x-\xi)| \, . \, |K_R(\xi)| \, dv_\xi = 0, \tag{1.1.8}$$

and it can be relaxed if we correspondingly tighten the assumptions on $K_R(\xi)$. For any measurable set A in E_k we can introduce the $L_p(A)$-norm, $p \geqq 1$,

$$\|f\|_p = \sup_{x \in E_k} \left(\int_A |f(x-\xi)|^p \, dv_\xi \right)^{1/p}, \tag{1.1.9}$$

and for $A = E_k$ this simply is

$$\left(\int_{E_k} |f(\xi)|^p \, dv_\xi \right)^{1/p}.$$

Also, if A is the set

$$T_k: -\tfrac{1}{2} \leqq \xi_j < \tfrac{1}{2}, \quad j = 1, \ldots, k, \tag{1.1.10}$$

and if $f(\xi_1, \ldots, \xi_k)$ is (multi)-periodic with period 1 in each variable, then the $L_p(T_k)$-norm is the L_p-norm of $f(x)$ over a fundamental domain of periodicity. Now, it follows from the Holder inequality

$$\left| \int_{T_k} g(\xi) \, dv_\xi \right| \leqq \left(\int_{T_k} |g(\xi)|^p \, dv_\xi \right)^{1/p} \left(\int_{T_k} 1^q \, dv_\xi \right)^{1/q},$$

that if, for a given $K_R(\xi)$, (1.1.8) holds for every function for which

$$\sup_{x \in E_k} \int_{T_k} |f(x-\xi)| \, dv_\xi < \infty, \tag{1.1.11}$$

then it also holds for every function with finite $L_p(E_k)$ or $L_p(T_k)$-norm. Now, (1.1.11) means that $|f(x)|$ becomes bounded after having been averaged over a T_k-neighbourhood of each point, and for such an $f(x)$, the integral

$$\int_{E_k} f(\xi_j) \, K(\xi_j) \, dv$$

is definable, whenever we have

$$\sum_{(m)} \sup_{\xi \in T_k} | K(\xi_1 + m_1, \ldots, \xi_k + m_k) | < \infty, \tag{1.1.12}$$

the summation extending over all lattice points $(m_1, \ldots, m_k)$, $m_j \gtreqless 0$.

Next, (1.1.12) holds in particular if we have for all $\xi \in E_k$

$$| K(\xi_1, \ldots, \xi_k) | \leqq \frac{C}{1 + | \xi |^{k+\rho}} \tag{1.1.13}$$

for some $C > 0$, no matter how large, and some $\rho > 0$, no matter how small. Also, if we form the special family (1.1.5), then the estimate

$$| K_R(\xi) | \leqq \frac{CR^k}{1 + R^{k+\rho} | \xi |^{k+\rho}}$$

implies

$$| K_R(\xi) | \leqq \frac{1}{R^\rho} \frac{C}{| \xi |^{k+\rho}}$$

for $| \xi | \geqq \delta$, $R \geqq 1$, and this secures relation (1.1.8) under the assumption (1.1.13). We do not claim that the mere condition (1.1.12) would secure (1.1.8), but it could be shown that the condition

$$\sum_{p=0}^{\infty} 2^{pk} \sup_{2^p \leqq | \xi | \leqq 2^{p+1}} | K(\xi_1, \ldots, \xi_k) | < \infty, \tag{1.1.14}$$

which falls between (1.1.12) and (1.1.13) would already suffice, and hence the following theorem:

THEOREM 1.1.2. *For a family of the form* (1.1.5), *if the kernel* $K(\xi)$ *satisfies* (1.1.13), *or only* (1.1.14), *then theorem* 1.1.1 *also applies if, globally,* $f(x)$ *has a finite norm* $L_p(E_k)$ *or only* $L_p(T_k)$.

If we put

$$K(\xi_1, \ldots, \xi_k) = \begin{cases} 1 & \text{for} \quad \xi \text{ in } T_k, \\ 0 & \text{for} \quad \xi \text{ not in } T_k, \end{cases} \tag{1.1.15}$$

then

$$s_{1/2h}(x) = \frac{1}{2h} \int_{-h}^{h} \cdots \int_{-h}^{h} f(x_1 + \xi_1, \ldots, x_k + \xi_k) \, dv_\xi, \tag{1.1.16}$$

and in particular for the simple exponential

$$\chi(\alpha; x) = e^{-2\pi i(\alpha, x)}, \quad (\alpha, x) \equiv \alpha_1 x_1 + \ldots + \alpha_k x_k \tag{1.1.17}$$

it is

$$\chi(\alpha; x) \cdot \prod_{j=1}^{k} \frac{\sin \pi h \alpha_j}{\pi h \alpha_j}. \tag{1.1.18}$$

The kernel (1.1.15) is a product kernel, in the sense that we have

$$K(\xi_1, \ldots, \xi_k) = K^0(\xi_1) \ldots K^0(\xi_k),$$

where $K^0(\xi)$ is a kernel in E_1.

Another product kernel is the (nonperiodic) *Fejer kernel*

$$\prod_{j=1}^{k}\left(\frac{\sin \pi\xi_j}{\pi\xi_j}\right)^2, \tag{1.1.19}$$

which, however, although it satisfies (1.1.12), does not satisfy (1.1.14) and thus could not be used in theorem 2. With a kernel $K^0(\xi)$ we may also form the multi-index kernel

$$K_{(R)}(\xi_1, \ldots, \xi_k) = K^0_{R_1}(\xi_1) \ldots K^0_{R_k}(\xi_k),$$

and most statements would be valid if $R_1, \ldots, R_k$ tend to ∞ independently of each other, but we will not pursue this possibility.

Of paramount importance is the *Gaussian kernel*

$$e^{-\pi(\xi_1^2+\cdots+\xi_k^2)} \equiv e^{-\pi|\xi|^2}, \tag{1.1.20}$$

which in addition to being a product kernel is also, antithetically, a *radial function*, meaning that there is a function $H(u)$ in $0 \leqq u < \infty$, such that

$$K(\xi_j) = H(|\xi|). \tag{1.1.21}$$

We will take as known the formula

$$\int_{-\infty}^{\infty} e^{-\pi\alpha^2 - 2\pi i\alpha x}\, d\alpha = e^{-\pi x^2}, \tag{1.1.22}$$

which implies

$$e^{-\pi(|\alpha|^2/R^2)} = R^k \int_{E_k} e^{-\pi R^2|\xi|^2 + 2\pi i(\alpha,\xi)}\, dv_\xi, \tag{1.1.23}$$

and this time we obtain for the function (1.1.17) the approximation

$$s_R(\alpha; x) = \chi(\alpha; x)\, e^{-\pi(|\alpha|^2/R^2)}. \tag{1.1.24}$$

For any radial kernel (1.1.21) it is profitable to introduce in

$$s_R(x) = \int_{E_k} f(x+\xi)\, H(R|\xi|)\, R^k\, dv_\xi$$

polar coordinates

$$\xi_j = t\eta_j, \quad 0 \leqq t \equiv |\xi| < \infty, \quad \eta_1^2 + \ldots + \eta_k^2 = 1,$$

in which case the volume element dv_ξ is the product of $t^{k-1}\,dt$ with the volume element $d\omega_\eta$ on

$$S_{k-1}\colon \eta_1^2 + \ldots + \eta_k^2 = 1, \tag{1.1.25}$$

the total volume of S_{k-1} being $\omega_{k-1} = \dfrac{2\pi^{(\frac{1}{2}k)}}{\Gamma(\frac{1}{2}k)}$. We then obtain

$$\begin{aligned} s_R(x) &= \omega_{k-1}\int_{t=0}^{\infty} f_x(t)\, H(Rt)\, R^k t^{k-1}\, dt \\ &= \omega_{k-1}\int_{t=0}^{\infty} f_x\left(\frac{t}{R}\right) H(t)\, t^{k-1}\, dt, \end{aligned} \tag{1.1.26}$$

where
$$f_x(t) \equiv \frac{1}{\omega_{k-1}} \int_{S_{k-1}} f(x_1+t\eta_1, \ldots, x_k+t\eta_k)\, d\omega_\eta \tag{1.1.27}$$

is the spherical average of our function at distance t from the given point x. By Fubini's theorem, $f_x(t)$ exists for almost all t, and we are always permitted to put $f_x(0)=f(x_j)$, and a glance at the term $I_1(R; x)$ in the proof to theorem 1.1.1 leads to the following conclusion:

THEOREM 1.1.3. *In theorems* 1.1.1 *and* 1.1.2, *if* $K(\xi_j)$ *is a radial function, then locally it suffices to assume that the spherical average* $f_x(t) \to f_x(0)$ *as* $t \to 0$, *which is a weaker assumption than continuity proper.*

For $k=2$ we have $\omega_1 = 2\pi$ and

$$f_x(t) = \frac{1}{2\pi} \int_0^{2\pi} f(x_1 + t\cos\theta, x_2 + t\sin\theta)\, d\theta.$$

However, for $k=1$ we have $\omega_0 = 2$ and $f_x(t) = \frac{1}{2}[f(x+t)+f(x-t)]$, and radiality means evenness, $K(-\xi)=K(\xi)$. For $k=1$, a function is even if it is invariant with respect to the (then only nontrivial) orthogonal transformation $\xi' = -\xi$ which leaves the origin fixed. Now, for $k \geqq 2$, radiality means invariance with respect to the entire group of such orthogonal transformations and $f_x(t)$ was an average over this group. However, if $K(\xi_j)$ is invariant with respect to a subgroup only, then the function $f(x+\xi)$ may be averaged correspondingly. Thus if $K(\xi_1, \ldots, \xi_k)$ is even with respect to each ξ_j separately, then it suffices to assume in theorems 1.1.1 and 1.1.2 that the averaged function

$$\frac{1}{2^k} \Sigma f(x_1 \pm \xi_1, \ldots, x_k \pm \xi_k)$$

shall be continuous at $\xi = 0$.

Turning for a moment to the smoothing operation (1.1.16) we note that by iterating it (or by some such procedure) we obtain the following result:

LEMMA 1.1.1. *A continuous function* $f(x_j)$ *in* E_k *which is* 0 *outside a compact set is a limit, uniformly in* E_k, *of suchlike functions each of which is of class* $C^{(r)}$ (*that is, has continuous partial derivatives of order* $\leqq r$) *for any fixed* r.

The conclusion also holds more precisely in the class C^∞, but of this we will not make use in primary contexts.

1.2. Translation functions

In E_k we take a family $\mathscr{F}$ of functions $\{f(x)\}$ with the following properties: (i) it is a group of addition, meaning that if $f, g \in \mathscr{F}$ then $f-g \in \mathscr{F}$; (ii) it is invariant with regard to translations, that is, if $f(x) \in \mathscr{F}$, and if for any $u=(u_1, \ldots, u_k)$ in E_k, we define

$$f^u(x) \equiv f(x_1+u_1, \ldots, x_k+u_k),$$

then $f^u(x) \in \mathscr{F}$; and (iii) it is endowed with a norm $\|f\|$ such that

$$\|0\|=0, \quad 0 \leqq \|f\| < \infty, \tag{1.2.1}$$

$$\|-f\| = \|f\|, \quad \|f+g\| \leqq \|f\| + \|g\|, \tag{1.2.2}$$

and, what is important, this norm is invariant, that is,

$$\|f^u\| = \|f\|. \tag{1.2.3}$$

With any $f \in \mathscr{F}$ we associate a certain non-negative function in $E_k: (u_1, \ldots, u_k)$, namely, the function

$$\tau_f(u) = \|f^u - f\|, \tag{1.2.4}$$

and we call it the *translation function* of f. It has the following properties. First,

$$\tau_f(0) = 0 \tag{1.2.5}$$

by (1.2.1). Next, due to

$$\|f^u - f\| \leqq \|f^u\| + \|-f\| = \|f\| + \|f\| = 2\,\|f\| = M,$$

we have

$$0 \leqq \tau_f(u) \leqq M = 2\,\|f\|. \tag{1.2.6}$$

Next, we have

$$\|f^{u+v} - f^v\| = \|(f^u - f)^v\| = \|f^u - f\|,$$

the last by (1.2.3), and on putting $v = -u$ we obtain

$$\tau_f(-u) = \tau_f(u). \tag{1.2.7}$$

Next, we have $\quad \|f^{u+v} - f\| \leqq \|f^{u+v} - f^u\| + \|f^u - f\|$

and hence

$$\tau_f(u+v) \leqq \tau_f(u) + \tau_f(v), \tag{1.2.8}$$

and finally for $f, g \in \mathscr{F}$ we have

$$\|f^u + g^u - f - g\| \leqq \|f^u - f\| + \|g^u - g\|$$

and hence

$$\tau_{f+g}(u) \leqq \tau_f(u) + \tau_g(u). \tag{1.2.9}$$

Now, (1.2.8) implies

$$\tau_f(u+v) - \tau_f(u) \leqq \tau_f(v), \tag{1.2.10}$$

but we also have

$$\tau_f(v) \leqq \tau_f(u) - \tau_f(u+v) = \tau_f(-u) - \tau_f(-u-v),$$

and if herein we replace $-u$ by $u+v$ we obtain

$$-\tau_f(v) \leqq \tau_f(u+v) - \tau_f(u). \tag{1.2.11}$$

Combining (1.2.10) and (1.2.11) and also using (1.2.5) we obtain

$$|\tau_f(u+v) - \tau_f(u)| \leqq \tau_f(v) = \tau_f(v) - \tau_f(0)$$

and hence the following conclusion:

LEMMA 1.2.1. *If a translation function is continuous at the origin it is uniformly continuous throughout.*

Next, by the use of (1.2.8) and (1.2.9) we now obtain by a familiar reasoning the following conclusion:

LEMMA 1.2.2. *If $\mathscr{F}_0$ is a dense (in norm) subset of $\mathscr{F}$ and if $\tau_f(u)$ is continuous in u for f in $\mathscr{F}_0$ it is continuous for f in $\mathscr{F}$.*

But if $\mathscr{F}$ is a normed vector space, more can be stated.

LEMMA 1.2.3. *If $\mathscr{F}$ is a normed vector space and if $\tau_f(u)$ is continuous in u for a set $\mathscr{F}_0$ whose linear combinations are dense in $\mathscr{F}$, then it is continuous for all of $\mathscr{F}$.*

This follows from

$$\tau_{c_1 f_1 + \ldots + c_n f_n}(u) \leqq |c_1| \tau_{f_1}(u) + \ldots + |c_n| \tau_{f_n}(u).$$

Now, for all finite multi-intervals

$$I_{ab}: a_j \leqq x_j < b_j, \quad j = 1, \ldots, k \tag{1.2.12}$$

we introduce the 'characteristic functions'

$$\omega_{ab}(x) = \begin{cases} 1 & \text{for} \quad x \text{ in } I_{ab}, \\ 0 & \text{for} \quad x \text{ not in } I_{ab}, \end{cases} \tag{1.2.13}$$

and it is a basic fact of the Lebesgue theory that their linear combinations are for every $1 \leqq p < \infty$ dense in the $L_p(E_k)$-space with the norm

$$\left(\int_{E_k} |f(\xi)|^p dv_\xi \right)^{1/p}.$$

On the other hand, we have

$$\tau_{\omega_{ab}}(u) = \left(\int_{E_k} |\omega_{ab}(\xi+u) - \omega_{ab}(\xi)|^p dv_\xi \right)^{1/p}$$

$$= \left(\int_{E_k} |\omega_{a-u,b-u}(\xi) - \omega_{ab}(\xi)|^p dv_\xi \right)^{1/p},$$

and it is easy to verify that this tends to 0 as $|u| \to 0$. Similarly, if we introduce for the periodic functions

$$f(x_1 + m_1, \ldots, x_k + m_k) = f(x_1, \ldots, x_k)$$

the $L_p(T_k)$-norm

$$\left(\int_{T_k} |f(\xi)|^p dv_\xi\right)^{1/p} = \left(\int_{T_k} |f(x+\xi)|^p dv_\xi\right)^{1/p},$$

then linear combinations of periodic functions of the form (1.2.13) are again dense in norm. Hence the following conclusion:

THEOREM 1.2.1. *For functions in $L_p(E_k)$ and (periodic) functions f in $L_p(T_k)$, $1 \leqq p < \infty$, the translation functions are (bounded) and continuous.*

We note that the general norm as defined by formula (1.1.9) is invariant with respect to translations, but we do not at all claim that every function with a finite norm of this kind has a continuous $\tau_f(u)$. However, if we take any set of functions $\{\mathscr{F}'\}$ each of which is bounded and uniformly continuous, and then form their smallest Banach closure with respect to the norm for a set A of finite Lebesgue measure, then their translation functions are continuous. If we choose for $\{\mathscr{F}'\}$ the simple exponentials

$$e^{i(\lambda_1 x_1 + \cdots + \lambda_k x_k)},$$

and for A the set T_k, then the smallest closure is composed of the almost periodic functions of the Stepanoff class L_p, to which we will sometimes refer incidentally.

1.3. Approximation in norm

We will now state a certain proposition first in a general version heuristically and then in a specific version precisely.

THEOREM 1.3.1 (*heuristic*). *If $\tau_f(u)$ is continuous then the approximating function*

$$s_R(x) = \int_{E_k} f(x-\xi)\, K_R(\xi)\, dv_\xi \tag{1.3.1}$$

converges to $f(x)$ in norm: $\|s_R - f\| \to 0$ as $R \to \infty$.

Reasoning. For a finite discrete sum we have

$$\begin{aligned}\left\| \sum_m \gamma_m (f(x+\xi_m) - f(x)) \right\| &\leqq \sum_m |\gamma_m| \cdot \|f^{\xi_m} - f\| \\ &= \sum_m |\gamma_m|\, \tau_f(\xi_m),\end{aligned}$$

and this suggests for (1.3.1) the estimate

$$\begin{aligned}\|s_R - f\| &= \left\| \int_{E_k} (f(x-\xi) - f(x))\, K_R(\xi)\, dv_\xi \right\| \\ &\leqq \int_{E_k} \|f^\xi - f\| . |K_R(\xi)|\, dv_\xi = \int_{E_k} \tau_f(\xi) . |K_R(\xi)|\, dv_\xi.\end{aligned}$$

But the last term is the value of

$$\int_{E_k} (\tau_f(x-\xi)-\tau_f(x)) \,.\, |K_R(\xi)| \, dv_\xi$$

for $x = 0$, and for bounded continuous $\tau_f(\xi)$ this tends to 0 as $R \to \infty$ as in theorem 1.1.1.

Assume now specifically that $f(x)$ is in $L_1(E_k)$. The function

$$H(x, \xi) = f(x-\xi) \, K_R(\xi)$$

is measurable in (x, ξ), and we have

$$\begin{aligned}\int_{E_k} \left(\int_{E_k} |f(x-\xi)| \,.\, |K_R(\xi)| \, dv_\xi \right) dv_x &= \int_{E_k} \left(|K_R(\xi)| \,.\, \int_{E_k} |f(x-\xi)| \, dv_x \right) dv_\xi \\ &= \int_{E_k} |K_R(\xi)| \, dv_\xi \,.\, \|f\| = \|f\| \,.\, K_0,\end{aligned}$$

and since the last term is finite, it follows by Fubini's theorem that the integral (1.3.1) exists for almost all x and is an integrable function in x. This being so, we now obtain

$$\begin{aligned}\int_{E_k} |s_R(x)-f(x)| \, dv_x &\leqq \int_{E_k} \left(\int_{E_k} |f(x-\xi)-f(x)| \,.\, K_R(\xi) | \, dv_\xi \right) dv_x \\ &= \int_{E_k} \left(\int_{E_k} |f(x-\xi)-f(x)| \, dv_x \right) . \, |K_R(\xi)| \, dv_\xi \\ &= \int_{E_k} \tau_f(\xi) \, |K_R(\xi)| \, dv_\xi,\end{aligned}$$

and this time rigorously. This argument also works for (periodic) $f \in L_1(T_k)$, if we replace one of the two symbols E_k by T_k, and thus we obtain the following theorem, at first for $p = 1$:

THEOREM 1.3.2. *If $f(x)$ belongs to $L_p(E_k)$ or to periodic or Stepanoff almost periodic $L_p(T_k)$, then the integral* (1.3.1) *exists for almost all x, is a function of the same class with*

$$\|s_R\| \leqq \|f\| \,.\, K_0$$

and we have

$$\lim_{R \to \infty} \|s_R - f\| = 0. \tag{1.3.2}$$

For $p > 1$ it is necessary to apply the Holder–Minkowski inequality

$$\left(\int_A \left(\int_B H(x, \xi) dv_x \right)^p dv_\xi \right)^{1/p} \leqq \int_B \left(\int_A H(x, \xi)^p \, dv_x \right)^{1/p} dv_\xi,$$

for $H(x, \xi)$ being first $|f(x-\xi)| . |K_R(\xi)|$ and then

$$|f(x-\xi)-f(x)| . |K_R(\xi)|$$

and $B = E_k$ and $A = E_k$ or T_k.

1.4. Vector-valued functions

Theorem 1.1 on convergence at a point and theorem 1.3.2 on convergence in strong average can be brought together by a third theorem embracing them both.

We define in E_k a function $\tilde{f}(x_j)$ whose values $\tilde{f}$ are not ordinary complex numbers but more generally elements of a Banach space $\tilde{B}$ the norm of which will be denoted by $\| \; \|$. For $\tilde{f}(x)$ we employ the concept of (strong) measurability and (strong) integrability as introduced by this author (the so-called Bochner integral), and if $\tilde{f}(x)$ is bounded in norm

$$\| \tilde{f}(x) \| \leqq M, \quad x \in E_k, \tag{1.4.1}$$

then for numerical $K_R(\xi_j)$ in $L_1(E_k)$ there exist the approximating functions

$$\tilde{s}_R(x) = \int_{E_k} \tilde{f}(x-\xi) K_R(\xi) \, dv_\xi \tag{1.4.2}$$

as functions again with values in $\tilde{B}$. We have

$$\| \tilde{s}_R(x) - \tilde{f}(x) \| \leqq \int_{E_k} \| \tilde{f}(x-\xi) - \tilde{f}(x) \| . |K_R(\xi)| \, dv_\xi$$
$$= \int_{|\xi| \leqq \delta} + \int_{|\xi| > \delta},$$

and thus continuity in norm at a point x,

$$\lim_{|\xi| \to 0} \| \tilde{f}(x-\xi) - \tilde{f}(x) \| = 0 \tag{1.4.3}$$

implies convergence in norm

$$\lim_{R \to \infty} \| \tilde{s}_R(x) - \tilde{f}(x) \| = 0, \tag{1.4.4}$$

and hence the following conclusion:

THEOREM 1.4.1. *Theorems* 1.1.1, 1.1.2 *and* 1.1.3 *also apply to functions* $\tilde{f}(x)$ *with values in a Banach space* $\tilde{B}$.

We now take in E_k: $(y_1, \ldots, y_k)$ a family $\mathscr{F}$ as in section 1.2, assuming that it is a Banach space and an element $f(y)$ in $\mathscr{F}$ for which the translation function $\tau_f(x)$ is continuous. If now we denote by $\tilde{f}(x)$ what in 1·2 we denoted by f^τ, then this $\tilde{f}(x)$ is bounded and continuous in norm and thus falls under theorem 1.4.1. In a certain formal sense we can write in (1.4.2)

$$s_R(y+x) = \int_{E_k} f(y+x-\xi)\, K_R(\xi)\, d\xi_\xi,$$

where $s_R(y+x)$ is $\tilde{s}_R(x)$, and if our Banach norm is $L_p(E_k)$ or the periodic or almost periodic $L_p(T_k)$ then this is rigorously so, for almost all y, for any given fixed x, as can be realized by assuming first that $f(y)$ is a finitely valued function as in section 1.2 and then passing to a limit in norm. But if this is so then relation (1.4.4), if applied at $x = 0$, is simply $\| s_R(y) - f(y) \| \to 0$ in the sense of theorem 1.3.2, and thus theorem 1.3.2 appears likewise subsumed under theorem 1.4.1.

Now, the (heuristic) theorem 1.3.1 has also a (heuristic) converse to the effect that if $s_R(x)$ converges in norm to $f(x)$ then $\tau_f(u)$ is continuous. But in the specific version in which we will establish this rigorously, we will not start out from a point function at all but from a (more general) set function, prove for it a 'weak' approximation by $s_R(x)$, and then show that if this approximation is also a strong one then the set function is the indefinite integral of a point function, and $\tau_f(u)$ is continuous.

1.5. Additive set functions

We denote by $V(E_k)$ the vector space of set functions

$$F(A) = F^1(A) + iF^2(A)$$

which are defined and σ-additive on the σ-field of (ordinary) Borel sets in E_k, the norm being the supremum

$$\| F \| = \sup_{(A_\nu)} \Sigma_\nu \, | F(A_\nu) |$$

for all partitions into disjoint sets. If $F(A)$ is real and $\geqq 0$ then $\| F \| = F(E_k)$, and the subset of such elements in V will be denoted by V^+. Any F in V can be written as

$$F(A) = F_1(A) - F_2(A) + iF_3(A) - iF_4(A), \tag{1.5.1}$$

where $F_1, F_2, F_3, F_4 \in V^+$, and there exists a $G \in V^+$ such that $| F(A) | \leqq G(A)$ for all A. Also, $\| F \| \leqq \| G \| = G(E_k)$. Now, among all such $G(A)$ there exists a 'smallest' one, which we will sometimes denote by $\tilde{F}(A)$ and call the 'absolute value' of $F(A)$, and it is characterized by

$$| F(A) | \leqq \tilde{F}(A), \quad \| F \| = \| \tilde{F} \|. \tag{1.5.2}$$

We will say that F is zero on a set B if $\tilde{F}(B)=0$, and this is equivalent to stating that $F(B_0)=0$ for any subset B_0 of B. Also if we are given a σ-additive set function $F(A)$ only for the subsets of a set A^0, then there is another set function in V which coincides with F for $A \subset A^0$, and is zero on $E_k - A^0$.

If $f(x) \in L_1(E_k)$ then

$$F(A)=\int_A f(x)\,dv_x \tag{1.5.3}$$

defines an element of V, the function $f(x)$ being defined up to sets of measure zero. We will call $F(A)$ an (indefinite) integral of $f(x)$ and $f(x)$ a 'derivative' of $F(A)$. If F is the integral of $f(x)$ then $\tilde{F}$ is the integral of $|f(x)|$. The elements in V which are integrals are a closed subset of V and constitute by themselves a Banach space which we will denote by $AC(E_k)$, the letters AC standing for 'absolutely continuous', and the meaning of this is that F in V belongs to AC if and only if $F(A_0)=0$ whenever $v(A_0)=0$.

Any $F \in V^+$ defines a bounded Lebesgue measure on the Borel sets of E_k, and every bounded Baire function $b(x_j)$ is integrable, and the integral will be denoted somewhat ambiguously by

$$\int_{E_k} b(x_j)\,dF(x_j), \tag{1.5.4}$$

the employment of the symbol $F(x_j)$ instead of $F(A)$ being somewhat arbitrary on the whole. For any $F \in V$ we take any decomposition (1.5.1) and define (1.5.4) by

$$\int b\,dF_1 - \int b\,dF_2 + i\int b\,dF_3 - i\int b\,dF_4, \tag{1.5.5}$$

and this integral is uniquely defined and has many customary properties.

If $F \in AC$, then (1.5.4) has the same value as the ordinary integral $\int_{E_k} b(x_j) f(x_j)\,dv_x$, and the familiar estimates

$$\left|\int bf dv_x\right| \leqq \int |b|\,.\,|f|\,dv_x \leqq \sup_x |b(x)| \int |f(x)|\,dv_x$$

can be generalized to

$$\left|\int bdF\right| \leqq \int |b|\,.\,|dF| \leqq \sup_x |b(x)| \int |dF|,$$

where $|dF(x)| \equiv d\tilde{F}(x)$, with $\tilde{F}(A)$ being the absolute value of $F(A)$, and we also have

$$\int_{E_k} |dF| - \int_{E_k} d\tilde{F} = \tilde{F}(E_k) = \|F\|.$$

The following properties are of considerable importance:

LEMMA 1.5.1. *If F_1, $F_2 \in V$, then $F_1 \equiv F_2$, that is $F_1(A) = F_2(A)$ for all A, whenever*

$$\int_{E_k} c(x)\,dF_1(x) = \int_{E_1} c(x)\,dF_2(x) \tag{1.5.6}$$

for every continuous $c(x)$ which vanishes outside a compact set, and (by lemma 1.1.1) *it suffices to assume that $c(x)$ belongs to class $C^{(r)}$, for some fixed r.*

Furthermore, if we are given F in V and a measurable function $f_0(x)$ which is integrable (with regard to ordinary measure) over any compact set and if we have

$$\int_{E_k} c(x)\,dF(x) = \int_{E_k} c(x) f_0(x)\,dx$$

for all such $c(x)$, then $F \in AC$ and $f_0(x)$ is its derivative.

Next, if F_1, F_2 in V have the same value in all 'octants'

$$I_a\colon -\infty < x_j < a_j, \quad j = 1, \ldots, k, \tag{1.5.7}$$

then $F_1 \equiv F_2$. On denoting $F(I_a)$ by $F(a_1, \ldots, a_k)$ we obtain a certain point function $F(x_j)$ which is representative of $F(A)$, and it is this interpretation of the symbol $F(x_j)$ which may be read into the formula (1.5.4). We will simply put $F(x_j) \in V$, and we note, what is much used in the theory of probability, that a function $F(x_j)$ is in V^+ if and only if

$$F(x_1, \ldots, x_k) \leqq F(y_1, \ldots, y_k) \quad \text{for} \quad x_1 \leqq y_1, \ldots, x_k \leqq y_k, \tag{1.5.8}$$

and also
$$\sum_{t_j=0}^{1} (-1)^{t_1+\cdots+t_k} F(y_j + t_j(x_j - y_j)) \geqq 0,$$

$$F(x_1 - 0, \ldots, x_k - 0) = F(x_1, \ldots, x_k), \tag{1.5.9}$$

$$\lim F(x_1, \ldots, x_k) = 0 \quad \text{if} \quad x_j \to -\infty \tag{1.5.10}$$

for a single index j, and

$$F(+\infty, \ldots, +\infty) \equiv \|F\| < \infty. \tag{1.5.11}$$

We will now take two elements F, G in V and make statements about them which we will explicitly discuss only if they are both in V^+, relying on a decomposition (1.5.1) for both of them if they are not. If we interpret $F(A)$ as a measure in space E_k^ξ, and $G(B)$ as a measure in a space E_k^η, then on the Borel sets of the product space

$$E_{2k} = E_k^\xi \times E_k^\eta \tag{1.5.12}$$

there is a product measure $\mu(C)$ such that

$$\mu(A \times B) = F(A) \,.\, G(B). \tag{1.5.13}$$

The functions $x_j = \xi_j + \eta_j, j=1,\dots,k$ are Baire functions in E_{2k}, and thus with any octant (1.5.7) in E_k there is associable the Baire set

$$C_a: -\infty < \xi_j + \eta_j < a_j, \quad j=1,\dots,k \tag{1.5.14}$$

in E_{2k}. If now we put $H(a_1,\dots,a_k) = \mu(C_a)$, then this defines a point function in $H(x_j)$ in V as previously described; and thus there is an element $H(C)$ in V such that for every bounded Baire function $b(x_j)$ in E_k we have

$$\int_{Ek} b(x)\, d_x H(x) = \int_{Ek}\int_{Ek} b(\xi+\eta)\, d_\xi F(\xi)\, d_\eta G(\eta), \tag{1.5.15}$$

the second integral being taken as a double integral or repeated integral, indifferently. Hence the following statement:

THEOREM 1.5.1. *With any two elements F, G in $V(E_k)$ there is associable a third, $H = F * G$ (their convolution) such that* (1.5.15) *holds*

The properties

$$F*G = G*F \quad \text{and} \quad (F_1*F_2)*F_3 = F_1*(F_2*F_3)$$

are obvious, but others will be stated formally although taken as known.

THEOREM 1.5.2. *For $H = F*G$ we have*

$$H(A) = \int_{Ek} F(A-\eta)\, d_\eta G(\eta) \tag{1.5.16}$$

for every set A.

If F has a derivative f then H has a derivative h, and if f is a bounded Baire function then

$$h(x) = \int_{Ek} f(x-\eta)\, d_\eta G(\eta).$$

If all three functions have derivatives, then

$$h(x) = \int_{Ek} f(x-\eta)\, g(\eta)\, d\eta$$

for almost all x.

DEFINITION 1.5.1. A sequence of elements $\{F_n\}$ in V *will be called Bernoulli convergent* if their norms are jointly bounded and if there is an element F in V such that

$$\lim_{n\to\infty} \int_{Ek} c(x)\, dF_n(x) = \int_{Ek} c(x)\, dF(x) \tag{1.5.17}$$

for every bounded continuous $c(x)$.

It will be called *weakly convergent* if (1.5.17) holds for every continuous $c(x)$ which vanishes outside a compact set, and (by lemma 1.1.1) it may be even assumed that $c(x)$ belongs to $C^{(r)}$ for a fixed r.

Certain decisive properties will be taken as known, and they are being stated as they will be needed.

LEMMA 1.5.2. (i) *Any infinite sequence* $\{F_n\}$ *in* $V(E_k)$ *with* $\|F_n\| \leqq M$ *contains an infinite subsequence which is weakly convergent, and if the entire sequence is not weakly convergent then it contains two weakly convergent subsequences whose limits are not identically equal.*

(ii) *For* $F_n \in V^+(E_k)$, *if* $\{F_n\}$ *is weakly convergent to* F, *then*

$$\varliminf_{n\to\infty} F_n(E_k) \geqq F(E_k), \tag{1.5.18}$$

and equality holds if and only if the weak convergence is Bernoulli convergence.

(iii) *If* $\{F_n\}$ *is weakly convergent and if all* F_n *are zero on an open set* A_0 *then the limit is also zero on* A_0.

(iv) *If* $\{F_n\}$ *is Bernoulli convergent and if a function* $\chi(\alpha; x)$ *is bounded and continuous for* $(\alpha_1, \ldots, \alpha_k)$ *in* E_k *and* $(x_1, \ldots, x_k)$ *in* E_k *then the limit relations*

$$\lim_{n\to\infty} \int_{E_k} \chi(\alpha; x)\, dF_n(x) = \int_{E_k} \chi(\alpha; x)\, dF(x) \tag{1.5.19}$$

holds uniformly in every compact set $|\alpha| \leqq \alpha_0$, $0 < \alpha_0 < \infty$.

(v) *If* $F_n \to F$ *weakly, and* $\|F_n\| \leqslant M$, *then* (1.5.17) *holds for every continuous* $c(x)$ *'which is zero at infinity', that is, for which* $\lim_{|x|\to\infty} c(x) = 0$.

All this will be needed later on, and for the present we are stating a theorem on approximation.

THEOREM 1.5.3. (i) *If* $F \in V(E_k)$, *and* $\{K_R(\xi)\}$ *is as before, then*

$$S_R(A) = \int_{E_k} F(A - \xi)\, K_R(\xi)\, dv_\xi \tag{1.5.20}$$

is Bernoulli convergent to $F(A)$, *as* $R \to \infty$.

(ii) $S_R(A)$ *is absolutely continuous, and if* $K_R(\xi)$ *is a continuous function then its derivative is*

$$s_R(x) = \int_{E_k} K_R(x - \xi)\, d_\xi F(\xi) \equiv \int_{E_k} K_R(\xi)\, d_\xi F(x - \xi). \tag{1.5.21}$$

(iii) *If* $S_R(A)$ *converges in norm to* $F(A)$,

$$\lim_{R\to\infty} \|S_R - F\| = 0, \tag{1.5.22}$$

then F *is absolutely continuous.*

(iv) *Finally, if* $K(\xi)$ *is as in theorem* 1.1.2, *then the convergence of* $s_R(x)$ *at a point* x *is a local property, meaning that if* F *is zero in a neighborhood of* x *then* $s_R(x) \to 0$ *at the point.*

Proof. We have

$$\Sigma_\nu | S_R(A_\nu) | \leqq \int_{E_k} \Sigma_\nu | F(A_\nu - \xi) | . | K_R(\xi) | dv_\xi$$

$$\leqq \int_{E_k} \| F \| . | K_R(\xi) | dv_\xi,$$

and thus

$$\| S_R \| \leqq \| F \| . K_0, \tag{1.5.23}$$

and for a bounded Baire function $b(x)$ we have

$$\int_{E_k} b(x)\, d_x S_R(x) = \int_{E_k} \left(\int_{E_k} b(x)\, d_x F(x-\xi) \right) K_R(\xi)\, dv_\xi \tag{1.5.24}$$

by Fubini's theorem. Now, the inside integral can also be written as

$$\int_{E_k} b(x+\xi)\, d_x F(x) \equiv \beta(\xi),$$

and if $b(x)$ is continuous then $\beta(\xi)$ is (bounded) and continuous, and $\int_{E_k} \beta(\xi) K_R(\xi)\, dv_\xi$ therefore converges to $\beta(0) \equiv \int_{E_k} b(x)\, d_x F(x)$, which proves part (i) of the theorem. Part (ii) is taken directly from theorem 1.5.2, and part (iii) follows from the fact that AC is a closed subset of V; and part (iv) states that under the assumptions of theorem 1.1.2 we have

$$\lim_{R \to \infty} \int_{|\xi| \geqq \delta} | K_R(\xi) | . | d_\xi F(x-\xi) | = 0$$

for $F(x) \in V(E_k)$, which can be easily verified.

Next, for F in V we introduce the translated element

$$F^u(A) \equiv F(A+u),$$

and the translation function

$$\tau^F(u) = \| F^u - F \|,$$

which, if $F(A)$ has a derivative $f(x)$ is our previous $\tau_f(u)$.

THEOREM 1.5.4. *For* F *in* $V(E_k)$, *if* $\tau^F(u)$ *is continuous, then* $F \in AC$.

Proof. We have

$$\Sigma_\nu | S_R(A_\nu) - F(A_\nu) | \leqq \int_{E_k} \Sigma_\nu | F(A_\nu - \xi) - F(A_\nu) | . | K_R(\xi) | dv_\xi,$$

and hence $$\| S_R - F \| \leqq \int_{E_k} \tau^F(\xi) . | K_R(\xi) | \, dv_\xi,$$

and for $\tau_F(\xi)$ continuous this tends to 0 as $R \to \infty$. Now apply part (iii) of theorem 1.5.3.

Theorem 1.5.4 also holds for a periodic function $F \in V(T_k)$ which we will introduce next, but we want to point out that it also holds on compact groups G, as we have shown elsewhere; that is, if we take a σ-additive set function $F(A)$ on the Borel sets in G and introduce the right translation function $\tau^F(u) = \| F(A) - F(uA) \|$ say, then $F(A)$ is absolutely continuous with regard to the Haar measure $v(A)$ if (and only if) $\tau^F(u)$ is continuous in u. An extension to noncommutative locally compact groups would be of some interest.

Furthermore, if a function $f(x)$ in $(-\infty, \infty)$ is integrable over every finite interval and if

$$\lim_{T \to \infty} \left(\frac{1}{T} \int_T^{T+a} + \int_{-T}^{-T+a} | f(x) | \, dx \right) = 0,$$

$-\infty < a < \infty$, then the function

$$\tau(u) = \overline{\lim_{T \to \infty}} \frac{1}{T} \int_{-T}^{T} | f(x+u) - f(x) | \, dx,$$

if finite, has all the properties of a translation function, and it would be of some interest to study the implications of the assumption that $\tau(u)$ is continuous without being identically zero, and the study ought to be extended to more general means of the form

$$\overline{\lim_{T \to \infty}} \frac{1}{T^p} \int_{-T}^{T} | f(x+u) - f(x) | \, dx,$$

say, $p > 0$.

1.6. Periodic additive set functions

Periodic set functions $F(A)$, like all other periodic functions, are to be thought of as being first introduced on the multitorus $T_k: -\frac{1}{2} \leqq x_j < \frac{1}{2}$ viewed as a compact space, and they may then be transplanted onto the entire E_k, as covering space of T_k, by periodic repetition

$$F(A+m) = F(A), \quad m = (m_1, \ldots, m_k),$$

so that they may be used in the formulas of theorem 1.5.3 which, as in the case of point functions, we wish to retain as they are, with integrations in them extending over the entire E_k.

If now we introduce the symbol $V(T_k)$ to designate the periodic σ-additive set functions (we will also employ the symbols $AC(T_k)$, $V^+(T_k)$, etc.), then we can first of all state as follows:

THEOREM 1.6.1. *Theorem 1.5.3 remains literally in force for $F \in V(T_k)$, provided Bernoulli convergence is now defined by*

$$\lim_{n\to\infty} \int_{T_k} c(x)\, dF_n(x) = \int_{T_k} c(x)\, dF(x) \tag{1.6.1}$$

for every continuous periodic function $c(x)$.

On the torus, due to its compactness, there is no difference between weak and Bernoulli convergence, and for instance any sequence in $V^+(T_k)$ for which $F_n(T_k) \leqq M$ contains a subsequence for which (1.6.1) holds. Thus, in this respect, the study of joint distribution functions of random positions on a closed wire is somewhat less sophisticated than for positions on the open infinite wire, on which the theory of statistics operates traditionally.

If $f(x)$ is periodic then for the integral

$$s_R(x) = \int_{E_k} f(x-\xi)\, K_R(\xi)\, dv_\xi \tag{1.6.2}$$

we can write formally

$$\sum_{(m)} \int_{T_k+(m)} f(x-\xi)\, K_R(\xi)\, dv_\xi = \sum_{(m)} \int_{T_k} f(x-\xi)\, K_R(\xi+m)\, dv_\xi,$$

and thus we have $$s_R(x) = \int_{T_k} f(x-\xi)\, \tilde{K}_R(\xi)\, dv_\xi, \tag{1.6.3}$$

where we have put $$\tilde{K}_R(\xi) = \sum_{(m)} K_R(\xi+m). \tag{1.6.4}$$

Now, by Lebesgue theory we have

$$\sum_{(m)} \int_{T_k} |K_R(\xi+m)|\, dv_\xi = \int_{T_k} \sum_{(m)} |K_R(\xi+m)|\, dv_\xi = \int_{E_k} |K_R(\xi)|\, dv_\xi,$$

and since, by assumption on $K_R(\xi)$, the last number is finite, the entire reasoning is rigorous in the following sense. The sum (1.6.4) is absolutely majorizedly convergent at almost all points x in T_k, and, as can be easily seen, in every compact subset of E_k, and the resulting sum function is independent of the order of the terms. Therefore $\tilde{K}_R(\xi_j)$ is a periodic element of $L_1(T_k)$ as seen from

$$\tilde{K}_R(\xi+p) = \sum_{(m)} K_R(\xi+m+p) = \sum_{(m)-(p)} K_R(\xi+m) = \tilde{K}_R(\xi),$$

and it has the following properties:

$$\int_{T_k} |(\tilde{K}_R(\xi)| \, dv_\xi \leqq K_0, \tag{1.6.5}$$

$$\int_{T_k} \tilde{K}_R(\xi) \, dv_\xi = 1, \tag{1.6.6}$$

$$\lim_{R\to\infty} \int_{|\xi|\geqq\delta,\, \xi\in T_k} \tilde{K}_R(\xi) \, dv_\xi = 0. \tag{1.6.7}$$

Furthermore, for $f \in L_p(T_k)$, (1.6.2) has indeed the value (1.6.3) a.e., but if we start from some kernels $\tilde{K}_R(\xi)$ on T_k with the properties stated, and if we introduce the approximating sums (1.6.3) and for $F \in V^+$ the sums

$$S_R(A) = \int_{T_k} F(A-\xi) \, \tilde{K}_R(\xi) \, dv_\xi, \tag{1.6.8}$$

then most of the previous theorems can be established likewise.

THEOREM 1.6.2. *For periodic point and set functions, theorems* 1.1.1, 1.3.2 *and* 1.5.3 *can also be established for the partial sums* (1.6.3) and (1.6.8).

As a rule we will represent periodic functions by the previous integrals over E_k, which are formally the same as for nonperiodic functions, but at one stage theorem 1.6.2 will be made use of, and for this utilization of it we are going to supplement it by a lemma in which R will have integer values n only.

LEMMA 1.6.1. *The (periodic) Fejer kernel*

$$\begin{aligned}\tilde{K}_n(\xi_j) &= \prod_{j=1}^{k} \frac{(\sin n\pi\xi_j)^2}{n(\sin \pi\xi_j)^2} \\ &\equiv \sum_{m_j=-n}^{n} \prod_{j=1}^{k} \left(1 - \frac{|m_j|}{n}\right) e^{2\pi i m_j \xi_j} \\ &= \sum_{(m)} \lambda_n(m) \, e^{2\pi i (m,\xi)},\end{aligned} \tag{1.6.9}$$

where $\quad \lambda_n(m) = \prod_{j=1}^{k} \left(1 - \dfrac{|m_j|}{n}\right) \quad$ if $\quad |m_1| \leqq n, \ldots, |m_k| \leqq n,$

and $\quad \lambda_n(m) = 0 \quad$ *if for some* j *we have* $|m_j| > n$,

has the properties (1.6.5), (1.6.6), (1.6.7) *needed.*

Proof. Since $\tilde{K}_R(\xi) \geqq 0$, (1.6.5) is implied by (1.6.6). Now we have $K_n(\xi_j) = H_n(\xi_1) \ldots H_n(\xi_k)$, where

$$H_n(\xi) = \frac{(\sin n\pi\xi)^2}{n(\sin \pi\xi)^2}, \tag{1.6.10}$$

and for $\delta < |\xi| \leqq \frac{1}{2}$ we have

$$H_n(\xi) \leqq \frac{1}{n(\sin \pi\delta)^2} \leqq \frac{A(\delta)}{n}, \tag{1.6.11}$$

so that *a fortiori*

$$\lim_{n\to\infty} \int_{\delta<|\xi|\leqq\frac{1}{2}} H_n(\xi)\, d\xi = 0. \tag{1.6.12}$$

We also have

$$\int_{-\frac{1}{2}}^{\frac{1}{2}} H_n(\xi)\, d\xi = 1, \tag{1.6.13}$$

and if in (1.6.7) we replace the exterior of the sphere $|\xi| < \delta$ by the exterior of the cube $|\xi_j| < \delta$, $j = 1, \ldots, k$, as we may, then it suffices to estimate a certain number of terms of the form

$$\int_{\xi_1=\delta}^{\frac{1}{2}} \int_{\xi_2=-\frac{1}{2}}^{\frac{1}{2}} \ldots \int_{\xi_k=-\frac{1}{2}}^{\frac{1}{2}} H_n(\xi_1)\, H_n(\xi_2) \ldots H_n(\xi_k)\, d\xi_1 \ldots d\xi_k.$$

Now, by (1.6.13) this reduces to $\int_{\xi_1=\delta}^{\frac{1}{2}} H_n(\xi_1)\, d\xi_1$, and this tends indeed to 0 as $n \to \infty$ by (1.6.12).

It should be noted that the sharper estimate (1.6.11) does not imply the same estimate for the product kernel (1.6.9), and the latter would again not be eligible for application in a theorem analogous to (1.1.2) on T_k instead of on E_k.

Finally, we ought to mention that the periodic Fejer kernel (1.6.9) is linked to the nonperiodic Fejer kernel (1.1.19) by relation (1.6.4) as could be deduced from theorem 2.4.1.

CHAPTER 2

FOURIER EXPANSIONS

2.1. Fourier integrals

For the present we will introduce Fourier integrals only for functions $L_1(E_k)$ and more generally $V(E_k)$, but not for $L_p(E_k)$, $p>1$, except that we will summarize statements on Plancherel transforms for functions $L_2(E_k)$ whose theory we will take as known. However, when introducing Fourier series, we will envisage L_p-classes in general because it takes very little effort to do so.

If $F \in V(E_k)$, we can introduce the Fourier transform

$$\phi(\alpha) \equiv \phi^F(\alpha_j) = \int_{E_k} e^{-2\pi i(\alpha, x)}\, dF(x), \tag{2.1.1}$$

and it is trivial that it is a bounded function,

$$\| \phi(\alpha) \| \leqq \| F \|, \tag{2.1.2}$$

and it is also very easy to see that it is uniformly continuous for (α_j) in E_k. If F has a derivative f we will also write

$$\phi(\alpha) = \phi_f(\alpha_j) = \int_{E_k} e^{-2\pi i(\alpha, x)} f(x)\, dv_x. \tag{2.1.3}$$

THEOREM 2.1.1. *If F, $G \in V(E_k)$ and $H = F * G$, then*

$$\phi^H(\alpha_j) = \phi^F(\alpha_j) \,.\, \phi^G(\alpha_j), \tag{2.1.4}$$

that is, the transform of a convolution is the product of the transforms.

In particular, if f, $g \in L_1$, then

$$\phi_h(\alpha) = \phi_f(\alpha)\, \phi_g(\alpha), \tag{2.1.5}$$

where

$$h(x) = \int_{E_k} f(x-y)\, g(y)\, dv_y, \tag{2.1.6}$$

the integral existing almost everywhere.

Proof. Put $b(x) = e^{-2\pi i(\alpha, x)}$ in formula (1.5.15).

THEOREM 2.1.2. *If F, $G \in V$, and*

$$\phi(\alpha_j) = \int e^{-2\pi i(\alpha, \xi)}\, dF(\xi), \quad \psi(\xi_j) = \int e^{2\pi i(\xi, \alpha)}\, dG(\alpha), \tag{2.1.7}$$

then

$$\int \phi(\alpha)\, e^{2\pi i(\alpha, x)}\, dG(\alpha) = \int \psi(x-\xi)\, dF(\xi). \tag{2.1.8}$$

Proof. Since

$$\left|\iint e^{2\pi i(\alpha, x-\xi)}\, dG(\alpha)\, dF(\xi)\right| \leqq \iint |\, dG(\alpha)\,| . |\, dF(\xi)\,| = \|G\| . \|F\| < \infty,$$

it follows by Fubini's theorem that the integral

$$\iint e^{2\pi i(\alpha, x-\xi)}\, dG(\alpha)\, dF(\xi)$$

exists, and relation (2.1.8) expresses the equality of the two repeated integrals by which it can be evaluated.

In the next theorem a function $\delta(\alpha_j)$ will be called a *convergence factor* if

$$\int_{E_k} |\, \delta(\alpha_j)\,|\, dv_\alpha < \infty, \tag{2.1.9}$$

and if for its (anti)-transform

$$K(\xi_j) = \int_{E_k} e^{2\pi i(\xi, \alpha)}\, \delta(\alpha)\, dv_\alpha$$

we have

$$\int_E |\, K(\xi_j)\,|\, dv_\xi < \infty, \quad \int_{E_k} K(\xi_j)\, dv_\xi = 1. \tag{2.1.10}$$

Now, the corresponding transform of $\delta(\alpha_j/R)$ is $R^k K(R\xi_1, \ldots, R\xi_k)$, and if in theorem 2.1.2 we put for $G(A)$ the indefinite integral of $\delta(\alpha_j)$, then the theorems of chapter 1 imply as follows:

THEOREM 2.1.3. *The integrals* (2.1.3) *and* (2.1.1) *can be inverted to*

$$f(x) \sim \int_{E_k} e^{2\pi i(x, \alpha)}\, \phi_f(\alpha)\, dv_\alpha \tag{2.1.11}$$

and

$$F(A) = \int_A dv_x \left[\int_{E_k} e^{2\pi i(x, \alpha)}\, \phi^F(\alpha)\, dv_\alpha\right] \tag{2.1.12}$$

in the following sense:

(i) *For any convergence factor* $\delta(\alpha)$ *the approximating function*

$$s_R(x) = \int_{E_k} \delta\left(\frac{\alpha}{R}\right) e^{2\pi i(x, \alpha)}\, \phi_f(\alpha)\, dv_\alpha \tag{2.1.13}$$

converges in norm to $f(x)$,

$$\lim_{R\to\infty} \int_{E_k} |\, f(x) - s_R(x)\,|\, dv_x = 0, \tag{2.1.14}$$

and for the approximating function

$$s_R(x) = \int_{E_k} \delta\left(\frac{\alpha}{R}\right) e^{2\pi i(x, \alpha)}\, \psi^F(\alpha)\, dv_\alpha \tag{2.1.15}$$

the indefinite integral $S_R(A)=\int_A s_R(x)\,dv_x$ *is Bernoulli convergent to* $F(A)$.

(ii) *For* $f(x)$ *bounded, the convergence of* $s_R(x)$ *at a point* x *is a local property, and for any* $f \in L_1$, *or even* $F \in V$, *it is so if*

$$|K(\xi)| \leqq C(1+(\xi)^{k+\rho})^{-1};$$

and in either case, $s_R(x) \to f(x)$ *if* $f(x)$ *is continuous at* x, *and for a radial* $K(\xi)$ *only the spherical average of* $f(\xi)$ *around* x *need be continuous at* x.

For us the dominant convergence factor will be $e^{-\pi|\alpha|^2}$ (and not the Abelian factor $e^{-|\alpha|}$) and we are going to utilize the corresponding sums

$$s_R(x)=\int_{E_k} e^{-\pi(|\alpha|^2/R^2)}e^{2\pi i(x,\alpha)}\phi^F(\alpha)\,dv_\alpha \tag{2.1.16}$$

for several purposes. If F_1, $F_2 \in V$ have equal transforms, then for $F=F_1-F_2$ we have $\phi^F(\alpha)=0$, and hence $s_R(x)=0$. But the Bernoulli limit of a null function is a null function, and thus part (i) of theorem 2.1.3 implies the following uniqueness theorem:

THEOREM 2.1.4. *If* $\phi^{F_1}(\alpha)\equiv\phi^{F_2}(\alpha)$, *then* $F_1\equiv F_2$. *If* $\phi_{f_1}(\alpha)\equiv\phi_{f_2}(\alpha)$ *then* $f_1(x)=f_2(x)$ *a.e.*

Next, assume that for F in V it so happens that

$$\int_{E_k}|\phi^F(\alpha)|\,dv_\alpha<\infty. \tag{2.1.17}$$

Since $|e^{-\pi(|\alpha|^2/R^2)}e^{2\pi i(x,\alpha)}|\leqq 1$ and $e^{-\pi(|\alpha|^2/R^2)}\to 1$ as $R\to\infty$, the functions (2.1.16) are then convergent, locally uniformly, towards the function

$$f_0(x)=\int_{E_k} e^{2\pi i(x,\alpha)}\phi^F(\alpha)\,dv_\alpha, \tag{2.1.18}$$

and for any continuous $c(x)$ which vanishes outside a compact set we have

$$\int_{E_k} c(x)\,s_R(x)\,dv_\alpha \to \int_{E_k} c(x)f_0(x)\,dv_x.$$

However, again by theorem 2.1.3 we have

$$\int_{E_k} c(x)\,s_R(x)\,dv_\alpha \to \int_{E_k} c(x)\,dF(x),$$

and by the second half of lemma 1.5.1 we obtain the following conclusion:

THEOREM 2.1.5. *If for* F *in* V, *the transform* $\phi^F(\alpha)$ *is in* $L_1(E_k)$, *then*

$F(A)$ is the indefinite integral of the continuous function $f_0(x)$ as given by the inversion formula (2.1.18).

In particular, if $f \in L_1$ and $\phi_f(\alpha) \in L_1$, then f is a.e. equal to the continuous function

$$f_0(x) = \int_{E_k} e^{2\pi i(x,\alpha)} \phi_f(\alpha)\, dv_\alpha. \tag{2.1.19}$$

As an incidental application of this theorem we note that a convergence factor $\delta(\alpha)$ as previously introduced may be assumed to be continuous to start with, in which case it is given by

$$\delta(\alpha) = \int_{E_k} e^{-2\pi i(\alpha,\xi)} K(\xi_j)\, dv_\xi,$$

so that in particular $\delta(0) = 1$. This being so, we will now give the actual formal definition of a convergence factor.

DEFINITION 2.1.1. *A convergence factor $\delta(\alpha)$* as was used in theorem 2.1.3 is a function with the following properties: $\delta(\alpha)$ is continuous and $\delta(0) = 1$, and $\delta(\alpha) \in L_1$, and its Fourier transform is in L_1.

2.2. Positive transforms. Plancherel transforms

THEOREM 2.2.1. *If for $f \in L_1$ we have $\phi_f(\alpha) \geqq 0$, and $|f(x)| \leqq M$ for $|x| \leqq t_0$ (that is, if f has a positive transform and is bounded in the neighborhood of the origin) then $\phi_f(\alpha) \in L_1$ and $f(x)$ is equal to the function $f_0(x)$* (see 2.1.19).

This theorem is obviously contained in the following one.

THEOREM 2.2.2. *If for $F \in V$ we have $\phi^F(\alpha) \geqq -\chi(\alpha)$, where $\chi(\alpha) \geqq 0$, $\chi(\alpha) \in L_1$, and if we have*

$$\left| \frac{1}{t^k} \int_{|\xi| \leqq t} dF(\xi) \right| \leqq M_1, \quad 0 < t < \infty, \tag{2.2.1}$$

or only

$$\left| R^k \int_{|\xi| \leqq 1} e^{-\pi R^2 |\xi|^2} dF(\xi) \right| \leqq M_2, \tag{2.2.2}$$

then $\phi^F(\alpha) \in L_1$, and $F(A)$ is the indefinite integral of the function $f_0(x)$ given by (2.1.18).

Proof. We note that due to $\| F \| < \infty$, (2.2.1) is satisfied automatically for $|t| \geqq 1$. Next, (2.2.2) is indeed at least as general as (2.2.1), this being an abelian theorem, since, if we put $G(t) = \int_{|\xi| \leqq t} dF(\xi)$, so that

$G(t)=O(t^k)$, we have

$$R^k\int_{|\xi|\leqq 1} e^{-\pi R^2|\xi|^2}\,dF(\xi)=R^k\int_0^1 e^{-\pi R^2t^2}\,dG(t)$$

$$=R^k e^{-\pi R^2}G(1)+2\pi R^{k+2}\int_0^1 te^{-\pi R^2t^2}O(t)^k\,dt$$

$$=O(1)+O(1)\int_0^\infty R^{k+2}e^{-\pi R^2t^2}t^{k+1}\,dt=O(1)$$

as $R\to\infty$. On the other hand, if it is known that $F(A)\geqq 0$, then (2.2.2) is not more general than (2.2.1), this being the classical Tauberian theorem, and (2.2.1) is the 'natural' way of putting the condition then.

Now, for the function (2.1.16) we have

$$s_R(0)=R^k\int_{E_k} e^{-\pi R^2|\xi|^2}\,dF(\xi)=\int_{|t|\leqq 1}+\int_{|t|>1},$$

but $\left|\int_{|t|>1}\right|\leqq R^k e^{-\pi R^2}\|F\|\to 0$, and therefore (2.2.2) is equivalent with

$$|s_R(0)|\leqq M_3.$$

Therefore

$$0\leqq\int_{E_k}(\phi(\alpha)+\chi(\alpha))\,e^{-\pi(|\alpha|^2/R^2)}\,dv_\alpha=s_R(0)+\int_{E_k}\chi(\alpha)\,e^{-\pi(|\alpha|^2/R^2)}\,dv_\alpha,$$

and because of the assumptions on $\chi(\alpha)$ this is $\leqq M_3+M_4=M_5$. Therefore, letting $R\to\infty$, we obtain

$$\int\phi(\alpha)+\chi(\alpha)\,dv_\alpha<\infty,\quad\text{and hence}\quad \phi^F(\alpha)\in L_1,$$

and now apply theorem 2.1.5.

Definition 2.2.1. (In the theory of probability) a *(joint) distribution function* $F(A)$ or $F(x_j)$ in E_k is an element $F\in V^+$ which is normalized by $\|F\|\equiv F(E_k)=1$, and its Fourier transform $\phi\ (\alpha_j)$ is called a *characteristic function.*

For such data we obtain the following special conclusion:

Theorem 2.2.3. *If for k random variables $X_1,\dots,X_k$ the joint distribution function $F(x_j)$ has a non-negative characteristic function and if*

$$\overline{\lim_{r=0}}\,\frac{1}{r^k}P\{|X_1|+\dots+|X_k|\leqq r\}<\infty,$$

then $F(A)$ is the integral of the function

$$f(x_j)=\int_{E_k} e^{-2\pi i(\alpha,x)}\phi(\alpha)\,dv_\alpha,$$

this expression converging absolutely.

If $\phi(\alpha)$ is the transform of f in $L_1(E_k)$, then $\phi(\alpha)\,\overline{\phi(\alpha)}$ is the transform of $g(x)=\int_{E_k} f(x+\xi)\overline{f(\xi)}\,dv_\xi$ (see theorem 2.1.1). If, by chance, f belongs also to $L_2(E_k)$, then by Schwarz's inequality we have

$$|g(x)-g(y)| \leqq \int |f(x+\xi)-f(y+\xi)|\cdot|f(\xi)|\,dv_\xi$$
$$\leqq \left(\int |f(x+\xi)-f(y+\xi)|^2 dv_\xi\right)^{\frac{1}{2}} \left(\int |f(\xi)|^2\right)^{\frac{1}{2}}$$
$$=\tau_f(x-y)\,.\,\|f\|,$$

where $\tau_f(u)$ is the translation function in the L_2-norm. Therefore, $g(x)$ is continuous, and by theorem 2.2.1 we have

$$\int f(x+\xi)\overline{f(\xi)}\,dv_\xi = \int \phi(\alpha)\,\overline{\phi(\alpha)}\,e^{2\pi i(x,\alpha)}\,dv_\alpha. \tag{2.2.3}$$

If f_1, f_2 are two such functions, then

$$\int f_1(x+\xi)\overline{f_2(\xi)}\,dv = \int \phi_1(\alpha)\,\overline{\phi_2(\alpha)}\,e^{2\pi i(x,\alpha)}\,dv_\alpha \tag{2.2.4}$$

and in particular
$$\int f_1(\xi)\overline{f_2(\xi)}\,dv_\xi = \int \phi_1(\alpha)\,\overline{\phi_2(\alpha)}\,dv_\alpha \tag{2.2.5}$$

$$\int |f(\xi)|^2 dv_\xi = \int |\phi(\alpha)|^2 dv_\alpha. \tag{2.2.6}$$

Now, by taking limits in L_2-norm the restriction that f, f_1, f_2 shall be not only in L_2 but also in L_1 can be eliminated and the following statements can be obtained:

THEOREM 2.2.4. *With each $f(x)$ in $L_2(E_k)$ there is associable another function $\phi(\alpha)=\phi_f(\alpha)$ likewise in $L_2(E_k)$, both determined a.e., such that* (i) *relations* (2.2.4), (2.2.5) *and* (2.2.6) *hold,* (ii) *each $\phi(\alpha)$ in L_2 is the transform of some f in L_2,* (iii) *for any Borel set A the inversion formulas hold:*

$$\int_A f(x)\,dv_x = \int_{E_k} \phi_f(\alpha)\left(\int_A e^{2\pi i(x,\alpha)}\,dv_x\right) dv_\alpha, \tag{2.2.7}$$

$$\int_{E_k} f(x)\left(\int_A e^{-2\pi i(\alpha,x)}\,dv_\alpha\right) dv_x = \int_A \phi_f(\alpha)\,dv_\alpha, \tag{2.2.8}$$

and (iv) *if $f(x)$ is also in L_1 then $\phi_f(\alpha)$ is the ordinary transform.*

If it also happens that the integral

$$g(x) = \int_{E_k} \phi_f(\alpha)\,e^{2\pi i(x,\alpha)}\,dv_\alpha$$

is boundedly convergent, say, and if we integrate both sides with respect to a set A and compare with (2.2.7), we obtain

$$\int_A f(x)\, dv_x = \int_A g(x)\, dv_x,$$

so that $f(x) = g(x)$ *a.e.*

2.3. Fourier series

We are now turning to periodic point and set functions. For any $F \in V(T_k)$ we are introducing the system of Fourier coefficients

$$\phi(m) \equiv \phi^F(m) = \int_{T_k} e^{-2\pi i(m,\xi)}\, dF(\xi)$$

for all lattice points $m = (m_1, \ldots, m_k)$, and if F has a derivative $f(x)$ then this is of course

$$\phi(m) \equiv \phi_f(m) = \int_{T_k} e^{-2\pi i(m,\xi)} f(\xi)\, dv_\xi. \tag{2.3.1}$$

As before, we now have $|\phi^F(m)| \leqq \|F\|$ and for $H = F * G$ the convolution rule $\phi^H(m) = \phi^F(m) . \phi^G(m)$, and as an analogue to theorem 2.1.2 we are making the following statement. Given $F \in V(T_k)$, if a function $\psi(\xi)$ is given by an absolutely convergent series

$$\psi(\xi) = \sum_{(m)} \gamma(m)\, e^{2\pi i(m,\xi)}, \quad \sum |\gamma(m)| < \infty,$$

then (by a direct substitution of this expansion) we obtain

$$\int_{T_k} \psi(x - \xi)\, dF(\xi) = \sum_{(m)} \gamma(m)\, \phi(m)\, e^{2\pi i(m,x)}. \tag{2.3.2}$$

If, in particular, we put for $\psi(x)$ the periodic Fejer kernel of lemma 1.6.1, then we obtain a sequence of approximating functions

$$s_n(x) = \int_{T_k} \tilde{K}_n(x - \xi)\, dF(\xi) = \sum_{(m)} \lambda_n(m)\, \phi(m)\, e^{2\pi i(m,x)},$$

where

(i) $$|\lambda_n(m)| \leqq 1, \quad \lim_{n\to\infty} \lambda_n(m) = 1 \tag{2.3.3}$$

and

(ii) for every n only a finite number of 'multipliers' $\lambda_n(m)$ are $\neq 0$. Hence, by theorem 1.6.2, the following conclusion:

THEOREM 2.3.1. (i) *Every $F \in V(T_k)$ is the weak limit of indefinite integrals of certain exponential polynomials, and any $f \in L_p$, $1 \leqq p < \infty$ is the limit in norm, and every continuous f is the uniform limit of* *exponential polynomials, and the polynomials arise from the Fourier series formally by the insertion of certain multipliers* $\lambda_n(m)$. (ii) *If* $F_1, F_2 \in V(T_k)$ *have the same Fourier series then they are equal.* (iii) *If* $\Sigma |\phi^F(m)| < \infty$, *then* F *is the integral of*

$$f(x) = \Sigma\phi(m)\, e^{2\pi i(m,x)}. \tag{2.3.4}$$

On the basis of this theorem we can now introduce kernels in general.

THEOREM 2.3.2. *If* $\delta(\alpha)$ *is a convergence factor* (definition 2.1.1) *for which also*

$$|K(\xi)| \leqq C(1+|\xi|^{k+\rho})^{-1} \tag{2.3.5}$$

then for any $F \in V(T_k)$, *the approximating function*

$$s_R(x) = \int_{E_k} K_R(x-\xi)\, dF(\xi) \tag{2.3.6}$$

has the Fourier series

$$s_R(x) \sim \sum_{(m)} \delta\left(\frac{m}{R}\right) \phi(m)\, e^{2\pi i(m,x)}. \tag{2.3.7}$$

Proof. We have already stated in section 1.1 that (2.3.5) implies

$$\sum_{(m)} \sup_{x,\xi \in T_k} |K_R(x_1-\xi_1+m_1, \ldots, x_k-\xi_k+m_k)| = M_R < \infty.$$

For fixed R if we start out with a *finite* sum

$$f^n(x) = \Sigma\phi^n(m)\, e^{2\pi i(m,x)} \tag{2.3.8}$$

we obtain

$$\begin{aligned} s_R^n(x) &\equiv \int_{E_k} K_R(x-\xi) f^n(\xi)\, dv \\ &= \sum_{(m)} \delta\left(\frac{m}{R}\right) \phi^n(m)\, e^{2\pi i(m,x)} \end{aligned} \tag{2.3.9}$$

by direct substitution. If now we take a general $f \in L_1(T_k)$ with a Fourier series

$$f(x) \sim \Sigma\phi(m)\, e^{2\pi i(m,x)},$$

then by the preceding theorem there is a sequence of polynomials (2.3.8) with $\|f^n - f\| \to 0$, so that automatically also $\lim_{n\to\infty} \phi^n(m) = \phi(m)$. Now

$$|s_R^n(x) - s_R(x)| \leqq \int_{E_k} |K_R(x-\xi)| \cdot |f^n(\xi) - f(\xi)|\, dv_\xi \leqq M_R \cdot \|f^n - f\|,$$

therefore the Fourier series of $s_R(x)$ is the formal limit of the series (2.3.9) and hence the conclusion.

For a general F in V, there is a sequence F^n of integrals of polynomials which converges Bernoulli to F on T_k, and this does not allow us to complete the last step of this reasoning quite so directly, but it suggests itself to make the following modification. We introduce the periodic kernel

$$\tilde{K}_R(\xi) = \sum_{(m)} K_R(\xi + m),$$

as in section 1·6, which, because of (2.3.5), is also continuous, and the uniform convergence of

$$s_R^n(x) = \int_{T_k} K_R(x-\xi)\, dF^n(\xi)$$

towards

$$s_R(x) = \int_{T_k} \tilde{K}_R(x-\xi)\, dF(\xi)$$

was formally treated in lemma 1.5.2.

However, the original reasoning on E_k itself had the advantage of applying step-by-step to the Stepanoff almost periodic functions as well; whereas, on the other hand, the description of a Stepanoff class of set functions $F(A)$ which would correspond precisely to the periodic class $V(T_k)$ has apparently not been given.

Theorem 2.3.2 has great technical advantages. For instance, if we put $x=0$ in the formula

$$R^k \int_{E_k} e^{-\pi R^2 |x-\xi|^2} dF(\xi) = \sum_{(m)} e^{-\pi(m^2/R^2)} \phi(m)\, e^{2\pi i (m, x)},$$

we obtain the following analogue to theorem 2.2.2:

Theorem 2.3.3. *If for $F \in V$ we have $\phi(m) \geqq -\chi(m)$, where $\chi(m) \geqq 0$, $\Sigma\chi(m) < \infty$, and if $F(\xi)$ satisfies the same boundedness condition at $x=0$ as in theorem 2.2.2, then $\Sigma\, |\, \phi(m)\, | < \infty$ and F is the indefinite integral of*

$$f(x) = \Sigma \phi(m)\, e^{2\pi i (m, x)}.$$

Finally, for $f \in L_2(T_k)$, this implies again

$$\int_{T_k} f(x+\xi)\overline{f(\xi)}\, dv_\xi = \sum_{(m)} |\, \phi(m)\, |^2 e^{2\pi i (m, x)},$$

and in particular

$$\int_{T_k} |f(\xi)\,|^2\, dv_\xi = \sum_{(m)} |\, \phi(m)\, |^2,$$

but this and the Riesz–Fischer theorem will be taken as known anyway.

2.4. Poisson summation formula

We have seen in section 1·6 that for $f(x)$ in $L_1(E_k)$ the series

$$\tilde{f}(x) = \sum_{(m)} f(x+m) \tag{2.4.1}$$

converges in $L_1(T_k)$-norm, and that for any continuous periodic $c(x)$ we have

$$\int_{T_k} \tilde{f}(x)\, c(x)\, dv_x = \int_{E_k} f(x)\, c(x)\, dv_x.$$

Hence the following conclusion:

THEOREM 2.4.1. *If $f(x) \in L_1(E_k)$, and*

$$\phi(\alpha) \equiv \phi_f(\alpha) = \int_{E_k} f(x) e^{-2\pi i(\alpha, x)}\, dv_x$$

is its Fourier transform, then we have

$$\sum_{(m)} f(x+m) \sim \sum_{(m)} \phi(m)\, e^{2\pi i(m, x)} \tag{2.4.2}$$

in the sense that the series on the right is the Fourier series of the function (2.4.1).

Thus in particular $\sum_{(m)} e^{-\pi |m/R|^2} \phi(m)\, e^{2\pi i(m, x)}$

converges in $L_1(T_k)$-norm to $\tilde{f}(x)$, and it converges literally to $f(x)$ at every point where $\tilde{f}(x)$ is continuous or at least where its spherical mean $\tilde{f}_x(t)$ is continuous.

For practical purposes the following special conclusion is important:

THEOREM 2.4.2. *If $f(x)$ in $L_1(E_k)$ is continuous and the series* (2.4.1) *is uniformly convergent and $\Sigma\, |\, \phi(m)\, | < \infty$, then relation* (2.4.2) *is a true equality at all points x.*

For any (y_j), the Fourier transform of $f(x_j)\, e^{-2\pi i(y, x)}$ is $\phi(\alpha_j + y_j)$, and thus (2.4.2) formally generalizes to

$$\sum_{(m)} f(m+x)\, e^{-2\pi i(m+x, y)} = \sum_{(m)} \phi(m+y)\, e^{2\pi i(m, x)}. \tag{2.4.3}$$

Furthermore, for any nonsingular affine transformation $x' = Tx$, $x'_p = \sum_{q=1}^{k} t_{pq} x_q$, we have

$$\phi(\alpha) = \int e^{-2\pi i(\alpha, x')} f(x')\, dv_{x'} = |\det T\,| \int e^{-2\pi i(\alpha, Tx)} f(Tx)\, dv_x. \tag{2.4.4}$$

But $(\alpha, Tx) = (T'\alpha, x)$, where T' is transposed to T, and thus

$$\phi(\alpha) = |\det T\,|\, \psi(T'\alpha),$$

where $\psi(\alpha)$ is the transform of $f(Tx)$. Therefore,

$$|\det T\,|\, \psi(\alpha) = \phi((T')^{-1} \alpha),$$

and hence we obtain the following statement:

THEOREM 2.4.3. *If $\phi(\alpha)$ is the transform of $f(x)$ then for any point x, y and any affine transformation T we have*

$$|\det T| \sum_{(m)} f(Tm+Tx)\, e^{-2\pi i(m+x,\, y)} = \sum_{(m)} \phi((T')^{-1}(m+y))\, e^{2\pi i(m,\, x)}. \quad (2.4.5)$$

The most renowned application of this is

$$\frac{1}{t^{\frac{1}{2}k}} \sum_{(m)} e^{-(\pi/t)|m+x|^2} = \sum_{(m)} e^{-\pi t|m|^2+2\pi i(m,\, x)} \quad (2.4.6)$$

and the full affine version of this is

$$\sum_{(m)} e^{-\pi Q(m+x)+2\pi i(m+x,\, y)} = (\det Q)^{-\frac{1}{2}} \sum_{(m)} e^{-\pi Q'(m+y)-2\pi i(m,\, x)}, \quad (2.4.7)$$

where $Q(\xi)$ is a nonsingular real symmetric quadratic form and $Q'(\xi)$ is its 'inverse'.

If we take as known the formula

$$t \int_{E_k} e^{-2\pi t|\alpha|+2\pi i(\alpha,\, x)}\, dv_\alpha = \frac{\pi^{-\frac{1}{2}(k+1)}\, \Gamma\{(\frac{1}{2}k+1)\}}{(t^2+|x|^2)^{\frac{1}{2}(k+1)}}$$

then we obtain

$$\frac{\Gamma\{\frac{1}{2}(k+1)\}}{\pi^{(\frac{1}{2}k+1)}} \sum_{(m)} \frac{t}{(t^2+|m+x|^2)^{\frac{1}{2}(k+1)}} = \sum_{(m)} e^{-2\pi t|m|+2\pi i(m,\, x)}$$

for any $k \geqq 1$, and especially

$$\frac{1}{\pi} \sum_{-\infty}^{\infty} \frac{t}{t^2+(m+x)^2} = \sum_{-\infty}^{\infty} e^{-2\pi t|m|+2\pi imx}$$

for $k=1$. However, in this case the right side can be 'summed', the sum being

$$\frac{1-e^{-4\pi t}}{1-2e^{-2\pi t}\cos 2\pi x+e^{-4\pi t}},$$

and on this point we will still comment.

Finally, we ought to mention the case of the 'discrete' L_p-space in E_k which arises by taking as underlying space the set of all lattice points $M_k=\{m\}$ in E_k and making it into a measure space by assigning the measure 1 to each point singly. The $L_p(M_k)$-space thus arising is the space of sequences $\{f(m)\}$ with

$$\|f\| = (\Sigma |f(m)|^p)^{1/p} < \infty,$$

and the analogue to the set functions $V(M_k)$ need not be analyzed separately, every such set function being absolutely continuous now. Any element f with finite L_1-norm

$$\Sigma|f(m)|<\infty$$

has a transform
$$\sum_{(m)} f(m)\, e^{-2\pi i(\alpha, m)} = \phi(\alpha)$$

which is continuous and periodic, the inversion formula being

$$f(m) = \int_{T_k} e^{2\pi i(m, \alpha)}\, \phi(\alpha)\, dv_\alpha.$$

Uniqueness, convolution, and the theorem on 'positive transforms' are trivial now, and the analogue to the Plancherel theorem for $L_2(M)$ is part of the Riesz–Fischer theorem already mentioned. The approximating sums for convergence factors are now

$$s_R(m) = R^k \sum_{(m)} K_R(m-n) f(n) = \int_{E_k} \delta\left(\frac{\alpha}{R}\right) \phi(\alpha)\, e^{2\pi i(m, \alpha)}\, dv_\alpha,$$

and, in particular, we have

$$\int_{E_k} e^{-\pi(|\alpha|^2/R^2)}\, \phi(\alpha)\, e^{2\pi i(m,\alpha)}\, dv_\alpha = R^k \sum_{(n)} e^{-\pi R^2 |m-n|^2} f(m),$$

which is worth stating perhaps, for the sake of analogy.

2.5. Summability. Heat and Laplace equations

Some of the findings of the preceding sections may be summarized as follows:

THEOREM 2.5.1. *Whether we are dealing with a Fourier integral or a Fourier series*

$$f(x) \sim \begin{cases} \displaystyle\int_{E_k} \phi(\alpha_j)\, e^{2\pi i(\alpha, x)}\, dv_\alpha \\ \displaystyle\sum_{(m)} \phi(m)\, e^{2\pi i(m, x)} \end{cases}$$

if we take a convergence factor $\delta(\alpha_j)$ in E_k with $\delta(0)=1$ and put

$$K(\xi_j) = \int_{E_k} e^{2\pi i(\xi, \alpha)}\, \delta(\alpha_j)\, dv_\alpha,$$

then the approximating functions

$$S_R(x) = \begin{cases} \displaystyle\int_{E_k} \delta\left(\frac{\alpha_j}{R}\right) \phi(\alpha_j)\, e^{2\pi i(\alpha, x)}\, dv_\alpha \\ \displaystyle\sum_{(m)} \delta\left(\frac{m_j}{R}\right) \phi(m_j)\, e^{2\pi i(m, x)} \end{cases}$$

exist and they have the 'identical' value

$$s_R(x) = R^k \int_{E_k} K(R(x_1 - \xi_1), \ldots, R(x_k - \xi_k)) f(\xi_j)\, dv_\xi.$$

This also applies to set functions $F(A)$ instead of point functions $f(x)$, and it also applies to Fourier series of Bohr and Stepanoff almost periodic functions.

As regards Cesaro–Riesz summability we have found that its natural 'spherical' version arises if we put

$$\delta(\alpha) = \begin{cases} (1 - |\alpha|^2)^\delta, & |\alpha| \leqq 1 \\ 0 & |\alpha| > 1 \end{cases} \tag{2.5.1}$$

[and *not* $\delta(\alpha) = (1 - |\alpha|)^\delta$ for $|\alpha| \leqq 1$, which for $\delta = 1$ and $k = 1$ happens to correspond to Fejer's kernel], and in this case we have

$$K(\xi_j) = K^\delta(\xi_j) = H^\delta(|\xi|),$$

where

$$H^\delta(u) = \frac{1}{\omega_{k-1}} \frac{J_{\delta + \frac{1}{2}k}(u)}{u^{\delta + \frac{1}{2}k}} \tag{2.5.2}$$

where $J_\nu(u)$ is the Bessel function. For large u this is $0((u)^{-\delta - \frac{1}{2}(k+1)})$, and thus for

$$\delta > \frac{k-1}{2} \tag{2.5.3}$$

we have $|K^\delta(\xi_j)| \leqq C(1 + |\xi|^{k+\rho})^{-1}$ with $\rho = \delta - \frac{1}{2}(k-1)$. Therefore, if δ exceeds $\frac{1}{2}(k-1)$ then *all* our theorems apply. The partial sum $s_R^\delta(x)$ exists at all points and, incidentally, has the value

$$s_R^\delta = R^{\frac{1}{2}k - \delta} \int_0^\infty f_x(u)\, u^{\frac{1}{2}k - \delta - 1} J_{\delta + \frac{1}{2}k}(uR)\, du, \tag{2.5.4}$$

and the behavior of $s_R^\delta(x)$ as $R \to \infty$ is a local property only. For the limiting exponent $\delta = \frac{1}{2}(k-1)$—we called it the 'critical' exponent—it was shown by Riemann for $k = 1$ that the localization property holds even then, although mere continuity of $f_x(u)$ at $u = 0$ no longer suffices. However, for $k \geqq 2$, the situation is more complex and more interesting, and although localization still holds for $f \in L_1(E_k)$ and even $f \in V(E_k)$, it no longer holds for $f \in L_1(T_k)$. In fact, as we have demonstrated, for each $k \geqq 1$ there is a periodic L_1-function which is 0 in a neighborhood of $x = 0$, and for which $s_R(0)$ is unbounded as $R \to \infty$. Furthermore, we showed by a subsequent argument that for $L_2(T_k)$ localization does hold, and we raised the problem for $L_p(T_k)$, $1 < p < 2$, and this problem is as yet unsolved.

We are now turning to the convergence factors $e^{-\pi|\alpha|^2}$, $e^{-2\pi|\alpha|}$ for an analysis of a different kind. In the case of the first factor we put $t = R^{-\frac{1}{2}}$, so that $0<t<\infty$, and $R\to\infty$ is to be replaced by $t\to 0$, and we denote the corresponding $s_R(x)$ by $s(t; x)$, so that by theorem 2.5.1 for the function

$$s(t;\, x)=\frac{1}{t^{\frac{1}{2}k}}\int_{E_k} e^{-(\pi/t)|x-\xi|^2}f(\xi)\,dv_\xi \tag{2.5.5}$$

we have the expansion

$$\int_{E_k} e^{-\pi t|\alpha|^2+2\pi i(\alpha,x)}\phi(\alpha)\,dv_\alpha \quad \text{or} \quad \sum_{(m)} e^{-\pi t|m|^2+2\pi i(m,x)}\phi(m) \tag{2.5.6}$$

respectively. In the second factor, however, we put $t=R^{-1}$, and for the function

$$s(t;\, x)=\frac{\Gamma\{\frac{1}{2}(k+1)\}}{\pi^{\frac{1}{2}(k+1)}}\int_{E_k}\frac{tf(\xi)\,dv_\xi}{(t^2+|x-\xi|^2)^{\frac{1}{2}(k+1)}}, \tag{2.5.7}$$

we then have

$$\int_{E_k} e^{-2\pi t|\alpha|+2\pi i(\alpha,x)}\phi(\alpha)\,dv_\alpha \quad \text{or} \quad \sum_{(m)} e^{-2\pi t|m|+2\pi i(m,x)}\phi(m) \tag{2.5.8}$$

respectively, and we recall that, for $k = 1$, for the last series we can also write

$$\int_{-\frac{1}{2}}^{\frac{1}{2}}\frac{(1-e^{-4\pi t})f(\xi)\,d\xi}{1-2\,e^{2\pi t}\cos 2\pi(x-\xi)+e^{-4\pi t}}, \tag{2.5.9}$$

alternately. Now, if in (2.5.6), say in the integral, we take the (formal) derivative with respect to t, then this introduces the factor

$$-\pi\,|\,\alpha\,|^2=-\pi(\alpha_1^2+\ldots+\alpha_k^2),$$

and, except for a constant, the same factor can be obtained by forming the negative Laplacean

$$\Delta_x\equiv-\left(\frac{\partial^2}{\partial x_1^2}+\ldots+\frac{\partial^2}{\partial x_k^2}\right) \tag{2.5.10}$$

with respect to $x_1, \ldots, x_k$. Also, since $|\,\phi(\alpha)\,|\leqq M$ it is very easy to verify that these differentiations are legitimate, and thus the function (2.5.5) is a solution of the heat equation

$$\frac{\partial s}{\partial t}=-\frac{1}{4\pi}\Delta s, \tag{2.5.11}$$

of which the given function $f(\xi)$ constitutes boundary values as $t\to 0$ in a manner described in our theorems. Next, in the case of the integral (2.5.8), one differentiation with respect to t introduces the factor $-2\pi\,|\,\alpha\,|$ which cannot be compensated by differentiations with respect to the variables x_j directly, but two differentiations with respect to t can be so compensated, and in fact we obtain the equation

$$\frac{\partial^2 s}{\partial t^2} = \Delta_x s \tag{2.5.12}$$

which may be interpreted as a Laplacean in $k+1$ variables

$$\frac{\partial^2 s}{\partial t^2} + \frac{\partial^2 s}{\partial x_1^2} + \ldots + \frac{\partial^2 s}{\partial x_k^2} = 0,$$

first of all. For $k=1$, formula (2.5.7) is the familiar solution for the boundary-value problem of harmonic functions in (y, t) in the half-plane $t > 0$, and formula (2.5.9) is the even more familiar solution in the unit circle $|z| < 1$, for the complex variable $z = e^{-2\pi(t+ix)}$.

However, from another approach, it is indicated to keep in (2.5.12) the variables (t, x) apart, and to take the operational square root on both sides of (2.5.12) writing it thus

$$\frac{\partial s}{\partial t} = -\Delta_x^{\frac{1}{2}} s, \tag{2.5.13}$$

as we will still discuss. This done, (2.5.13) falls under the general form of an operational equation

$$\frac{\partial f}{\partial t} = -A_x f, \tag{2.5.14}$$

in which A_x is a linear operator with certain positivity features as possessed by the Laplacean, primarily but not at all exclusively, and all such equations (2.5.14) will be interpreted as diffusion equations, as we will still see.

2.6. Theta relations with spherical harmonics

We now denote a Fourier transform in E_k by

$$g(y_j) = \int_{E_k} e^{-2\pi i(y,x)} f(x_j)\, dv_x, \tag{2.6.1}$$

and in both spaces we introduce spherical coordinates which we will denote by

$$x_j = |x| \xi_j, \quad \xi_1^2 + \ldots + \xi_k^2 = 1, \tag{2.6.2}$$

$$y_j = |y| \eta_j, \quad \eta_1^2 + \ldots + \eta_k^2 = 1. \tag{2.6.3}$$

Actually, in such coordinates the point of origin $(x_j) = (0)$ is a singular one, and, strictly taken, it must then be excepted in virtually all statements; however, we will do so explicitly only whenever its exceptional status is a material and not only a formal issue.

A function $f(x_j)$ is *radial* if there is a function $f_0(u)$ in $0<u<\infty$ such that $f(x_j)=f_0(|x|)$, and it turns out that for such a function the transform (2.6.1.) is likewise a radial one, $g(y_j)=g_0(|y|)$, and the functions of the radius are connected by a so-called Hankel transform

$$g_0(v)=\frac{2\pi}{v^k}\int_0^\infty f_0\left(\frac{u}{v}\right) J_{\frac{1}{2}k-1}(2\pi u)\, u^{\frac{1}{2}k}\, du$$

$$\equiv \frac{2\pi}{v^{\frac{1}{2}k-1}}\int_0^\infty f_0(u)\, J_{\frac{1}{2}k-1}(2\pi uv)\, u^{\frac{1}{2}k}\, du. \tag{2.6.4}$$

If, however, the 'angular' variables ξ_j do occur, then the dependence on them is best expressed by means of spherical harmonics which we are going to introduce now.

DEFINITION 2.6.1. (Somewhat ambiguously) a (*spherical harmonic*) is a function either in E_k, or on the unitsphere S_k: $\xi_1^2+\ldots+\xi_k^2=1$. If in E_k, a harmonic of order n, $P_n(x_j)$ is a homogeneous polynomial of order n which is a solution of the Laplacean

$$-\Delta_x P_n(x)\equiv\left(\frac{\partial^2}{\partial x_1^2}+\ldots+\frac{\partial^2}{\partial x_k^2}\right)P_n(x_j)=0,$$

and if on S_k it is a function $P_n(\xi_j)$ which is connected with a $P_n(x_j)$ by means of $P_n(\xi_j)=P_n(x_j/|x|)$ or, equivalently, $P_n(x_j)=|x|^n P_n(\xi_j)$, such a function satisfying the equation

$$\Delta_\xi P_n(\xi_j)=-n(n+k-2)P_n(\xi_j),$$

where Δ_ξ is the 'curvilinear' Laplacean on S_k.

LEMMA 2.6.1. *If $H_n(t;\xi_j)$ is continuous in $0<t<\infty$, $\xi_j\in S_k$, and is for each t an n-th harmonic in (ξ_j) and if the integral*

$$\int_0^\infty H_n(t;\xi_j)\,dt=H_n(\xi_j)$$

is absolutely uniformly convergent then $H_n(\xi_j)$ is again an n-th harmonic.

In fact, for given n, the nth harmonics are a vector space with finite basis, and a uniform limit of polynomials in real variables of the same degree is again a polynomial, and all (mixed) partial derivatives are uniformly convergent likewise.

LEMMA 2.6.2. *For any harmonic $P_n(\xi)$ the formula*

$$\int_{S_k} e^{-2\pi i Z(\eta,\xi)}P_n(\xi)\,d\omega_\xi=2\pi i^n P_n(y)\frac{J_{n+\frac{1}{2}k-1}(2\pi Z)}{Z^{\frac{1}{2}k-1}} \tag{2.6.5}$$

holds.

This formula is equivalent to a general version of the addition theorem for Bessel functions and will be taken as known.

THEOREM 2.6.1. *If $f(x_j) \in L_1(E_k)$ is of the form*

$$f(x_j) = \lambda(|x|)\, P_n(\xi), \tag{2.6.6}$$

then its transform (2.6.1) *is of the form*

$$g(y_j) = \mu(|y|)\, P_n(\eta), \tag{2.6.7}$$

where $P_n(\zeta)$ is the same harmonic in both formulas and

$$\mu(v) = \frac{2\pi i^n}{v^k} \int_0^\infty \lambda\left(\frac{u}{v}\right) J_{n+\frac{1}{2}k-1}(2\pi u)\, u^{\frac{1}{2}k}\, du. \tag{2.6.8}$$

Proof. For spherical coordinates $x_j = t\xi_j$ we have $dv_x = t^{k-1}\,dt\,d\omega_\xi$ and hence

$$g(y_j) = \int_0^\infty \lambda(t)\, t^{k-1} \left(\int_{S_k} e^{-2\pi i t |y| (\eta, \xi)} P_n(\xi)\, d\omega_\xi \right) dt,$$

and if we substitute (2.6.5) the assertion follows.

If we replace $\lambda(|x|)$ by $\lambda(|x|)\,|x|^n$ and $\mu(|y|)$ by $\mu(|y|)\,|y|^n$, we obtain the following variant:

THEOREM 2.6.2. *If $f(x_j)$ in $L_1(E_k)$ is of the form*

$$f(x_j) = \lambda(|x|)\, P_n(x_j),$$

then

$$g(y_j) = \mu(|y|)\, P_n(y_j)$$

and

$$\mu(v) = \frac{2\pi i^n}{v^{2n+k}} \int_0^\infty \lambda\left(\frac{u}{v}\right) J_{n+\frac{1}{2}k-1}(2\pi u)\, u^{n+\frac{1}{2}k}\, du, \tag{2.6.9}$$

and therefore, except for a factor i^n, the connection between $\lambda(u)$ and $\mu(v)$ in the case $(k; n)$ is the same as in the case $(2n+k; 0)$.

Now, in any dimension, the transform of $e^{-\pi |x|^2}$ is $e^{-\pi |y|^2}$, and hence the following conclusion:

THEOREM 2.6.3. *For any harmonic we have*

$$\int_{E_k} P_n(x_j)\, e^{-\pi |x|^2} e^{-2\pi i (y,x)}\, dv_x = i^n P_n(y_j)\, e^{-\pi |y|^2},$$

that is, $P_n(x_j)\, e^{-\pi |x|^2}$ is an eigenfunction, with eigenvalue i^n, for the Fourier transform operation in E_k.

If we now apply theorem 2.4.3 we obtain the following general theta relation:

THEOREM 2.6.4. *For any spherical harmonic*

$$P_r(x_1, \ldots, x_k), \quad r = 0, 1, 2, \ldots,$$

we have

$$\sum_{(m)} P_r(m_1, \ldots, m_k)\, e^{-\pi t(m_1^2+\cdots+m_k^2)} = i^r t^{-r-\frac{1}{2}k} \sum_{(m)} P_r(m_1, \ldots, m_k)\, e^{-(\pi/t)(m_1^2+\cdots+m_k^2)}, \tag{2.6.10}$$

for $0<t<\infty$. *More generally, for all* x, y *we have*

$$\sum_{(m)} R_r(m+x)\, e^{-\pi t Q(m+x)+2\pi i \Sigma_j (m_j+x_j) y_j}$$
$$= i^r t^{-r-\frac{1}{2}k} (\det Q)^{-\frac{1}{2}} \sum_{(m)} R'_r(m+y)\, e^{-(\pi/t)Q'(m+y)-2\pi i \Sigma_j (m_j+y_j) x_j}, \tag{2.6.11}$$

where $Q(m)$ *is any positive definite quadratic form in* $(m_1, \ldots, m_k)$ *and* $Q'(m)$ *is the inverse form, and where* $R_r(\eta)$ *arises from a harmonic polynomial* $P_r(\xi)$ *by a substitution* $\xi = A\eta$ *which transforms* $\sum m_j^2$ *into* $Q(m)$, *and where* $R'_r(\eta)$ *arises by the reciprocal transposed substitution* $\xi = (A^*)^{-1}\eta$.

For our next application the following fact will be taken as known:

LEMMA 2.6.3. *If* μ, ν *are any complex numbers with* $\mathrm{Re}(\mu+\nu)>0$, *then*

$$\lim_{\epsilon \downarrow 0} \int_0^\infty e^{-\epsilon t} J_\nu(t)\, t^{\mu-1}\, dt = \frac{2^{\mu-1}\Gamma\{\frac{1}{2}(\mu+\nu)\}}{\Gamma\{\frac{1}{2}(\nu-\mu)+1\}},$$

for all μ, ν. *If* $\nu-\mu+2=-m$, *in which case the gamma factor in the denominator is* ∞, *then the limit has the value* 0 *accordingly.*

Using this, if in theorem 2.6.1 we put $\lambda(u) = ((2\pi)^{\frac{1}{2}} u)^\gamma e^{-\epsilon u}$ then we obtain the significant relation

$$\lim_{\epsilon \downarrow 0} \int_{E_k} e^{-\epsilon |x|} ((2\pi)^{\frac{1}{2}} |x|)^\gamma P_n(\xi_j)\, e^{-2\pi i (y,x)}\, dv_x$$
$$= \frac{2^i P_n(\eta)}{((2\pi)^{\frac{1}{2}} |y|)^{k+\gamma}} \frac{2^{\frac{1}{2}k+\gamma}\Gamma\{\frac{1}{2}(\gamma+k+n)\}}{\Gamma\{\frac{1}{2}(n-\gamma)\}} \tag{2.6.12}$$

for $|y| \neq 0$. The constant γ may be complex, and if we want the integral to be absolutely convergent we must put $\mathrm{Re}\,\gamma > -k$. However, the relation also holds for $\mathrm{Re}\,\gamma > -k-n$, with the integral existing at the origin as a Cauchy principal value (spherical).

A remarkable situation arises for $\gamma = -\frac{1}{2}k$ which we will describe separately.

THEOREM 2.6.5. *For any harmonic* $P_n(\xi)$ *we have*

$$\lim_{\epsilon \downarrow 0} \int_{E_k} \frac{P_n(\xi_j)}{|x|^{\frac{1}{2}k}} e^{-\epsilon |x|} e^{-2\pi i (x,y)}\, dv_x = i^n \frac{P_n(\eta_j)}{|y|^{\frac{1}{2}k}},$$

and thus $P_n(\xi_j)\,|x|^{-\frac{1}{2}k}$ *is an eigenfunction of the Fourier transform operation, for the eigenvalue* i^n.

2.7. Expansions in spherical harmonics

We now introduce the coordinates (2.6.2), (2.6.3) into the general relation (2.6.1), and if we put

$$f(|x|;\xi_j)\equiv f(x_j),\quad g(|y|;\eta_j)\equiv g(y_j), \tag{2.7.1}$$

then this relation becomes

$$g(v;\eta_j)=\int_0^\infty\left(\int_{S_k} e^{-2\pi i v u(\eta,\xi)} f(u;\xi_j)\,d\omega_\xi\right)u^{k-1}\,du. \tag{2.7.2}$$

In order to utilize this we now take it as known that any function $\phi(\xi_j)$ on S_{k-1} which is continuous, or only belongs to $L_1(S_k)$, has a 'Fourier expansion' in spherical harmonics,

$$\phi(\xi_j)\sim\sum_{n=0}^\infty \phi_n(\xi_j), \tag{2.7.3}$$

by which it is uniquely determined, and we introduce these expansions for our functions (2.7.1), denoting them thus:

$$f(u;\xi_j)\sim\sum_{n=0}^\infty f_n(u;\xi_j), \tag{2.7.4}$$

$$g(u;\eta_j)\sim\sum_{n=0}^\infty g_n(v;\eta_j). \tag{2.7.5}$$

If now we substitute (2.7.4) in (2.7.2), and apply the uniqueness property, then we obtain the following conclusion, which can be established rigorously for any class of functions for which (2.6.1) can be defined:

THEOREM 2.7.1. *In spherical coordinates, the Fourier transformation (2.6.1) becomes the set of relations*

$$g_n(v;\eta_j)=\frac{2\pi i^n}{v^k}\int_0^\infty f_n\left(\frac{u}{v};\eta_j\right)J_{n+\frac{1}{2}k-1}(2\pi u)\,u^{\frac{1}{2}k}\,du,\quad n=0,1,2,\ldots. \tag{2.7.6}$$

Our first application of this will be a kind of converse to theorem 2.6.1.

THEOREM 2.7.2. *If a measurable function $\lambda(u)$ in $0<u<\infty$ is such that*

$$\int_0^\infty |\lambda(u)|\,e^{Au}du<\infty \tag{2.7.7}$$

for all $A>0$, and if

$$\int_0^\infty \lambda(u)\,u^\sigma\,du\neq 0 \tag{2.7.8}$$

for all $\sigma > 0$ (*thus, for instance, if* $\lambda(u) > 0$); *and if for a function* $\phi(\xi_j)$ *in* $L_1(S_k)$ *the function*

$$f(x_j) = \lambda(|x|)\,\phi(\xi_j) \tag{2.7.9}$$

is such that its Fourier transform $g(y_j)$ *is likewise of the form*

$$g(y_j) = \mu(|y|)\,\psi(\eta_j); \tag{2.7.10}$$

then $\phi(\xi_j)$ *must be a 'monomial' spherical harmonic* $P_n(\xi)$, *and* $g(y_j)$ *is as in theorem* 2.6.1.

Proof. If we introduce the expansion (2.7.3) then we have

$$f_n(u;\, \xi_j) = \lambda(u)\,\phi_n(\xi_j),$$

and by (2.7.6) we obtain

$$g_n(v;\, \eta_j) = \mu_n(v)\,\phi_n(\eta_j), \quad n = 0, 1, 2, \ldots, \tag{2.7.11}$$

where

$$\mu_n(v) = \frac{2\pi i^n}{v^{\frac{1}{2}k-1}} \int_0^\infty \lambda(u)\, J_{n+\frac{1}{2}k-1}(2\pi uv)\, u^{\frac{1}{2}k}\, du. \tag{2.7.12}$$

But by (2.7.10) we have alternate values

$$g_n(v;\, \eta_j) = \mu(v)\,\psi_n(\eta_j), \quad n = 0, 1, 2, \ldots, \tag{2.7.13}$$

and by comparing this with (2.7.11) we find that if for an index $n = 0, 1, 2, \ldots$ we do not have

$$\phi_n(\eta_j) \equiv 0, \tag{2.7.14}$$

then there is a constant c_n such that

$$\mu_n(v) = c_n\,\mu(v). \tag{2.7.15}$$

Now, by assumption (2.7.7) we may substitute the expansion

$$J_{n+\frac{1}{2}k-1}(2\pi uv) = \sum_{p=0}^{\infty} (-1)^p \frac{(\pi uv)^{n+\frac{1}{2}k-1+2p}}{p!\,\Gamma(n+\frac{1}{2}k+p)}$$

in the integral (2.7.12) and integrate term-by-term. This gives rise to a convergent expansion

$$\mu_n(v) = \sum_{p=0}^{\infty} a_{n,p}\, v^{n+2p},$$

and by assumptions (2.7.8) we have $a_{n,0} \neq 0$ so that the expansion begins with v^n effectively. Therefore, we cannot have (2.7.13) for two different indices n, and thus we must have (2.7.14) for all except one index n, and this is precisely what the theorem claims.

Now, the function $\lambda(u)=u^r e^{-\pi u^2}$, $r \geqq 0$, satisfies (2.7.7) and (2.7.8) and hence the following very.special inference:

THEOREM 2.7.3. *If we have*

$$\int_{E_k} P_r(x_j)\, e^{-\pi |x|^2 - 2\pi i(y,x)}\, dv_x = Q_s(y_j)\, e^{-\pi |y|^2}$$

for two homogeneous polynomials $P_r(x_j)$, $Q_s(y_j)$ of some degrees r, s, then $P_r(x_j)$ is a spherical harmonic, and we have $s=r$, $Q_s(y_j)=i^n P_r(y_j)$.

In particular, if for a homogeneous polynomial $P_r(x_j)$ we have

$$\int_{E_k} P_r(x_j)\, e^{-\pi |x|^2} e^{-2\pi i(y,x)}\, dv_x = cP_r(y_j)\, e^{-\pi |y|^2}, \tag{2.7.16}$$

then $P_r(x_j)$ is a harmonic and $c=i^n$.

We now recall that the (inhomogeneous) Hermite polynomials in one variable $H_n(x)$, $n=0,1,\ldots$, after replacing x by $(2\pi)^{\frac{1}{2}} x$, have the property

$$\int_{-\infty}^{\infty} H_n(x)\, e^{-\pi x^2} e^{-2\pi i y x}\, dx = i^n H_n(y)\, e^{-\pi y^2}.$$

However, any polynomial $P_n(x_j)$ can be expanded into a finite series

$$\sum_{n_1+\ldots+n_k \leqq n} a_{n_1 \ldots n_k} H_{n_1}(x_1) \ldots H_{n_k}(x_k), \tag{2.7.17}$$

and uniquely so. Therefore, the transform of $P_n(x_j)\, e^{-\pi |x|^2}$ is

$$Q_n(y_j)\, e^{-\pi |\eta|^2},$$

where $$Q_n(y_j) = \sum_{n_1+\ldots+n_k \leqq n} (i)^{n_1+\cdots+n_k} a_{n_1 \ldots n_k} H_{n_1}(x_1) \ldots H_{n_k}(x_k),$$

and since $P_n(x_j)$ is a homogeneous harmonic if and only if

$$Q_n(y_j) = i^n P_n(y_j),$$

we obtain the following theorem:

THEOREM 2.7.4. *A homogeneous polynomial $P_n(x_j)$ is a spherical harmonic if and only if, on representing it by an expansion (2.7.17), only such coefficients $a_{n_1 \ldots n_k}$ are $\neq 0$ for which*

$$n_1 + \ldots + n_k = n - 4r, \quad r = 0, 1, 2, \ldots.$$

Turning now to a somewhat different topic we note that relation (2.6.9) implies as follows:

THEOREM 2.7.5. *Subject to assumptions, if we are given on S_k a* *function* $\phi(\xi_j)$ *with the expansion* (2.7.3), *and if for some* γ *with* $\operatorname{Re}\gamma > -k$ *we consider on* S_k *the function with the expansion*

$$\psi^{\gamma}(\xi_j) = 2^{\frac{1}{2}k+\gamma} \sum_{0}^{\infty} \frac{i^n \Gamma\{\frac{1}{2}(\gamma+k+n)\}}{\Gamma\{\frac{1}{2}(n-\gamma)\}} \phi_n(\xi), \tag{2.7.18}$$

then the function $\quad f(x_j) = ((2\pi)^{\frac{1}{2}} \,|\, x \,|)^{\gamma} \phi(\xi_j)$

has for $(x_j) \neq (0)$ *the transform*

$$g(y_j) = ((2\pi)^{\frac{1}{2}} \,|\, y \,|)^{-\gamma-k} \psi^{\gamma}(\eta_j),$$

in the sense that we have

$$g(y_j) = \lim_{\epsilon \downarrow 0} \int_{E_k} e^{-\epsilon |x|} f(x_j)\, e^{-2\pi i(x,y)}\, dv_x,$$

the limit existing.

Now, as for specific assumptions that are sufficient, we quote the following theorem:

THEOREM 2.7.6. *For given* γ, *if* $2n$ *is an even integer for which* $2n > \operatorname{Re}\gamma + \frac{3}{2}k - 1$ *and if* $\phi(\xi_j)$ *belongs to differentiability class* $C^{(2n)}$ *on* S_k, *then the function* (2.7.18) *exists and is continuous and the conclusions of theorem* 2.7.5 *hold true.*

Also, if $q^0 \geqq 0$ *is an integer for which* $2n > \operatorname{Re}\gamma + \frac{3}{2}k - 1 + q^0$, *then* $\psi^{\gamma}(\xi_j) \in C^{(q^0)}$, *and thus if* $\phi(\xi_j) \in C^{(\infty)}$ *then also* $\psi^{\gamma}(\xi_j) \in C^{(\infty)}$, *for* $\operatorname{Re}\gamma > -k$.

Finally, if $\phi(\xi_j)$ *is (real) analytic on* S_k *in its 'natural' local coordinates, then* $\psi^{\gamma}(\xi_j)$ *is also analytic.*

The last-made statement on analyticity is non-obvious. The theorem applies in particular to

$$g(y_j; s) = \lim_{\epsilon \downarrow 0} \int_{E_k} e^{-\epsilon |x|} \frac{e^{-2\pi i(x,y)}\, dv_x}{(T_{2h}(x_j))^s}, \tag{2.7.19}$$

where $T_{2h}(x_j)$ is a homogeneous polynomial

$$\sum_{n_1+\dots+n_k=2h} a_{n_1\dots n_k} x_1^{n_1} \dots x_k^{n_k},$$

which is >0 for $|\, x \,| \neq 0$, and s is a complex number for which

$$\operatorname{Re} s < \frac{k}{2h}. \tag{2.7.20}$$

It suggests itself that the function $g(y; s)$ might also be holomorphic (that is, complex-analytic) in the parameter s in the half-plane (2.7.20), and this is not only correct but easy to prove. Altogether we will have in the next section a rather general theorem from which we will deduce at the end of section 2.9 that the function (2.7.19) is holomorphic in s and $C^{(\infty)}$ in (y_j), for $(y_j) \neq 0$. But the actual analyticity in the real variables y_j is a much more specific proposition and the proof will not be reproduced here.

2.8. Zeta integrals

DEFINITION 2.8.1. We say that a function $f(x_j; s)$ is an *adjusted Zeta kernel* if it has the following structure:

(i) It is defined for (x_j) in E_k, and for complex s in a certain domain D.

(ii) It is 0 in a fixed neighborhood $|x| < x^0$ of the origin in E_k.

(iii) It is $C^{(\infty)}$ in the variables x_j, and for any combination of integers

$$p_1 \geqq 0, \ldots, p_k \geqq 0 \tag{2.8.1}$$

the function
$$D^{p_1 \cdots p_k} f(x; s) \equiv \frac{\partial^{p_1 + \cdots + p_k} f(x_j; s)}{\partial x_1^{p_1} \ldots \partial x_k^{p_k}} \tag{2.8.2}$$

is continuous in $(x; s)$ and holomorphic in s, and what is decisive

(iv) corresponding to each s^0 in D there exists a neighborhood N of s^0 and a real number $\gamma = \gamma(N)$, $\gamma \gtreqless 0$, such that for each combination (2.8.1) we have

$$\frac{\partial^{p_1 + \cdots + p_k} f(x; s)}{\partial x_1^{p_1} \ldots \partial x_k^{p_k}} = 0(|x|^{\gamma - (p_1 + \cdots + p_k)}), \tag{2.8.3}$$

as $|x| \to \infty$, uniformly for s in N.

DEFINITION 2.8.2. In the present context a *convergence factor* is a function $\delta(t)$ of the following description: (i) it is defined and continuous in $0 \leqq t < \infty$, and $\delta(0) = 1$; (ii) $\delta(t)$ is $C^{(\infty)}$ in $0 < t < \infty$; (iii) we have

$$\frac{d^p \delta(t)}{dt^p} = O(t^{-q}) \quad \text{as} \quad t \to \infty \tag{2.8.4}$$

for every $p \geqq 0$, $q \geqq 0$, so that in particular

$$\delta(t) = O(t^{-q}) \quad \text{as} \quad t \to \infty \tag{2.8.5}$$

for every $q > 0$; and (iv) we have

$$\frac{d^p \delta(t)}{dt^p} = o(t^{-p}) \quad \text{as} \quad t \to 0 \tag{2.8.6}$$

for every $p > 0$ (but not $p = 0$).

The function $\delta(t) = e^{-t\rho}$ is such a convergence factor for every $\rho > 0$.

Definitions 2.8.1 and 2.8.2 are such that for any $\epsilon > 0$ the product

$$f_\epsilon(x_j; s) = \delta(\epsilon |x|) f(x_j; s) \tag{2.8.7}$$

is again an adjusted Zeta kernel; and, in consequence of (2.8.3) and (2.8.5), this new function and each mixed partial derivative of it is small at infinity in such a manner that for the transform

$$g_\epsilon(y_j; s) = \int_{E_k} e^{-2\pi i(x,y)} f_\epsilon(x_j; s)\, dv_x \tag{2.8.8}$$

we obtain the relation

$$(2\pi i y_1)^{p_1} \dots (2\pi i y_k)^{p_k} g_\epsilon(y; s) = \int_{E_k} e^{-2\pi i(x,y)} D^{p_1 \cdots p_k} f_\epsilon(x; s)\, dv_x \tag{2.8.9}$$

for any integers (2.8.1), by means of partial integrations which are readily justified. But now we claim as follows:

THEOREM 2.8.1. *If N and $\gamma = \gamma(N)$ are fixed, then for*

$$p_1 + \dots + p_k > \gamma + k, \tag{2.8.10}$$

we have

$$\lim_{\epsilon \downarrow 0} \int_{E_k} e^{-2\pi i(x,y)} D^{p_1 \cdots p_k} f_\epsilon(x; s)\, dv_x = \int_{E_k} e^{-2\pi i(x,y)} D^{p_1 \cdots p_k} f(x; s)\, dv_x, \tag{2.8.11}$$

uniformly in the neighborhood of every point (y, s).

Proof. If we write $D^{(r)}$ for $D^{r_1 \cdots r_k}$ and put $r = r_1 + \dots + r_k$, then we have

$$D^{(p)} f_\epsilon(x; s) = \delta(\epsilon\,|\,x\,|)\, D^{(p)} f(x; s) + \sum_{\substack{r+s=p \\ r>0}} D^{(r)} \delta(\epsilon\,|\,x\,|)\, D^{(s)} f(x; s). \tag{2.8.12}$$

Now, by (2.8.3), we have

$$D^{(p)} f(x; s) = O\,(\,|\,x\,|^{\gamma - p}), \quad |\,x\,| \to \infty,$$

and by (2.8.10) this function thus belongs to $L_1(E_k)$, uniformly in N. Also, $\delta(\epsilon\,|x\,|)$ converges boundedly to 1 as $\epsilon \to 0$, and thus our theorem will follow if we verify

$$\lim_{\epsilon \downarrow 0} \int_{E_k} |\, D^{(r)} \delta(\epsilon\,|\,x\,|)\,| \,.\, |\, D^{(s)} f(x; s)\,|\, dv_x = 0 \tag{2.8.13}$$

uniformly in s, for $r + s = p$, $r > 0$. Now, since $f(x; s) = 0$ for $|\,x\,| < x^0$ it is not hard to verify, by the use of (2.8.3), that the integral (2.8.13) is majorized by

$$\int_{x^0}^{\infty} \left| \frac{d^r}{dt^r} \delta(\epsilon t) \right| .\, t^{\gamma - s + k - 1}\, dt \equiv \epsilon^{p - \gamma - k} \int_{\epsilon x^0}^{\infty} |\, \delta^{(r)}(t)\,|\, t^{\gamma - s + k - 1}\, dt.$$

However, by assumption (2.8.10) the quantity $\rho - p - \gamma - k$ is > 0, and thus $\epsilon^{p - \gamma + k} \int_{\eta}^{\infty}$ tends to 0, as $\epsilon \to 0$, for any fixed η, no matter how small. Now, if $r>0$, then, in $\epsilon x^0 \leqq t \leqq \eta$, $|\delta^{(r)}(t)|$ is $\leqq \sigma(\eta) t^{-r}$, where $\sigma(\eta) \to 0$ as $\eta \to 0$, and thus we only have to verify that

$$\epsilon^{p-\gamma-k} \int_{\epsilon x^0}^{1} t^{\gamma-r-s+k-1}\, dt$$

stays bounded as $\epsilon \to 0$. But, due to $r+s=p$, this is a trivial fact, and hence the theorem.

The theorem implies in particular that for

$$2n > \gamma + k \tag{2.8.14}$$

the function
$$(y_1^2 + \dots + y_k^2)^n g_\epsilon(y; s) \equiv |y|^{2n} g_\epsilon(y; s) \tag{2.8.15}$$

has a limit, as $\epsilon \to 0$, uniformly locally in (y, s), and this limit is independent of the special factor $\delta(t)$ employed. However, we can divide the function (2.8.15) by the factor $|y|^{2n}$ whenever $|y| \neq 0$, and we thus obtain part (i) of the following theorem:

THEOREM 2.8.2. (i) *For any adjusted Zeta kernel the transform*

$$g(y_j; s) \sim \int_{E_k} e^{-2\pi i(x, y)} f(x_j; s)\, dv_x \tag{2.8.16}$$

exists for $|y| \neq 0$, *as a limit*

$$g(y_j; s) = \lim_{\epsilon \downarrow 0} g_\epsilon(y_j; s), \tag{2.8.17}$$

uniformly in the neighborhood of every point (y, s), *and the limit is independent of the particular convergence factor used in* (2.8.7).

(ii) *The function* $g(y_j; s)$ *is holomorphic in* s *and* $C^{(\infty)}$ *in* (y_j), *and the derivations with respect to the variables* y_j *can be performed under the integral sign in* (2.8.16).

(iii) *We have*
$$g_\epsilon(y_j; s) = O(|y|^{-q}), \quad \text{as} \quad |y| \to \infty \tag{2.8.18}$$

for every $q>0$, *uniformly say in* $0<\epsilon<1$ *and in the neighborhood of every point* s.

Proof. Ad (ii). The holomorphy in s follows from the fact, which is a consequence of Weierstrass's limit theorem, that a definite integral $\int h(x; s)\, dv_x$ is holomorphic in s, whenever $h(x; s)$ is holomorphic in s and the integral is approximable by finite Riemann sums locally uniformly in s. Next, for given $\epsilon > 0$, relation (2.8.8) implies

$$\frac{\partial g_\epsilon(y_j; s)}{\partial y_1} = \int_{E_k} e^{-2\pi i(x, y)} \delta(\epsilon |x|)(-2\pi i x_1) f(x_j; s)\, dv_x, \tag{2.8.19}$$

say. But $x_1 f(x_j;\, s)$ is again an adjusted Zeta kernel, and thus, by part (i), the function (2.8.19) must have a limit, as $\epsilon \to 0$, locally uniformly, and it now follows that this limit must be

$$\frac{\partial g_\epsilon(y_j;\, s)}{\partial y_1},$$

this derivative existing. We can now iterate this procedure, and the assertion follows.

Ad (iii). The function (2.8.15) has, except for a constant, the value

$$\int_{E_k} e^{-2\pi i(x,\, y)}\, \Delta_x^n f(x_j;\, s)\, dv_x,$$

where Δ is the Laplacean. It follows from our proof to theorem 2.8.1 that, for $2n > \gamma + k$, this function is bounded in y_j, uniformly for $0 < \epsilon < 1$ and s in N, but n can be chosen arbitrarily large whence the conclusion.

2.9. Zeta series

The unnatural restriction incorporated in definition 2.8.1 that $f(x;\, s)$ shall be 0 in a neighborhood of $x = 0$ was meant to be only transitory, and we are now going to allow, to the contrary, that it shall even be undefined and singular at $x = 0$.

Definition 2.9.1. We say that a function $f(x_j;\, s)$ is a *(proper) Zeta kernel* if it is defined for x in E_k except at the origin $x = 0$ and for s in a domain D, and if it has properties (iii) and (iv) of definition 2.8.1.

Theorem 2.9.1. (i) *If $f(x_j;\, s)$ is a Zeta kernel then the corresponding 'Zeta function'*

$$\zeta(y_j;\, s) = \sum_{(m)}{}' e^{-2\pi i(m,\, y)} f(m_j;\, s) \tag{2.9.1}$$

exists for $y \neq 0$ on the torus T_k: $-\frac{1}{2} \leqq y_j < \frac{1}{2}$ and for s in D as a limit

$$\lim_{\epsilon \downarrow 0} \sum_{(m)}{}' e^{-2\pi i(m,\, y)}\, \delta(\epsilon \mid m \mid) f(m_j;\, s) \tag{2.9.2}$$

for any convergence factor $\delta(t)$, and has a value independent of it, uniformly in the neighborhood of every point (y, s), and it is $C^{(\infty)}$ in y and holomorphic in s.

(ii) *Also, corresponding to $y = 0$, the limit*

$$\lim_{\epsilon \downarrow 0} \left[\sum_{(m)}{}' \delta(\epsilon \mid m \mid) f(m_j;\, s) - \int_{\mid x \mid \geqq 1} \delta(\epsilon \mid x \mid) f(x_j;\, s)\, dv_x \right] \tag{2.9.3}$$

exists, continuously for s in D, and is holomorphic there.

Proof. Ad (i). We take it as known that it is possible to construct a continuous function $\chi(t)$ in $0 \leqq t < \infty$ which is $C^{(\infty)}$ in $0 < t < \infty$ and for which we have $\chi(t) = 0$ for $0 \leqq t \leqq \frac{1}{3}$ and $\chi(t) = 1$ for $\frac{2}{3} \leqq \chi(t) < \infty$. If now, with the given function $f(x_j;\, s)$, we construct the new function

$$\tilde{f}(x_j;\, s) = \begin{cases} \chi(|\, x\,|) f(x_j;\, s), & |\, x\,| = 0, \\ 0, & x = 0, \end{cases}$$

then $\tilde{f}(x_j;\, s)$ is on the one hand an adjusted Zeta kernel, and on the other hand we have $\tilde{f}(x_j;\, s) = f(x_j;\, s)$ for $|\, x\,| \geqq \frac{2}{3}$ so that in particular we have

$$\sum_{(m)}{}' e^{-2\pi i (m,\, y)} f_\epsilon(m_j;\, s) = \sum_{(m)} e^{-2\pi i (m,\, y)} \tilde{f}_\epsilon(m_j;\, s).$$

However, by theorem 2.4.1 we have

$$\sum_{(m)} g_\epsilon(m_j + y_j) = \sum e^{-2\pi i (m,\, y)} \tilde{f}_\epsilon(m_j;\, s), \tag{2.9.4}$$

where g_ϵ is the Fourier transform of $\tilde{f}$, and if we apply all properties stated for g_ϵ in theorem 2.8.2 then our assertion follows.

Ad (ii). For $y = 0$ the term $g_\epsilon(0)$ need not have a limit as $\epsilon \to 0$. However, if we write

$$\sum_{(m)}{}' g_\epsilon(m_j) = \sum_{(m)}{}' f_\epsilon(m_j;\, s) - g_\epsilon(0),$$

then the left side does have a limit, and a holomorphic one too, again by theorem 2.8.2. Now

$$g_\epsilon(0) = \int_{E_k} f_\epsilon(x_j;\, s)\, dv_x,$$

and if we write for this

$$\int_{|\, x\,| \geqq 1} f_c(x_j;\, s)\, dv_x + \int_{\frac{1}{3} \leqq |\, x\,| \leqq 1} \tilde{f}_\epsilon(x_j;\, s)\, dv_x,$$

then the second integral is holomorphic in s for s in D, so that we need subtract only the first integral, and this was done so in formula (2.9.3).

We now take a homogeneous polynomial $T_{2h}(x_j) > 0$, for $x \neq 0$, as in section 2.7, and with it we set up any (nonhomogeneous) polynomial

$$T(x_j) = T_{2h}(x_j) + (\text{lower powers}),$$

and we assume that we have $T(x_j) \neq 0$ for $x \neq 0$. For $k \geqq 3$ it then follows by monodromy that we can define $(T(x_j))^s$ as a holomorphic function in s for $x \neq 0$ and all complex s, and for $k = 2$ we make this into an additional assumption (which is needed only if $T(x_j)$ is nonhomogeneous). We take some further polynomial $Q_r(x_j)$, say homogeneous of degree r, and if we put $f(x_j; s) = Q_r(x_j)\,T(x_j)^{-s}$, then this is a Zeta kernel with D being the entire s-plane. Therefore, theorem 2.9.1 implies immediately parts (i) and (ii) of the following theorem:

THEOREM 2.9.2. (i) *The function*

$$\zeta(y_j; s) = \lim_{\epsilon \downarrow 0} \sum_{(m)}{}' e^{-2\pi i(y, m)} \delta(\epsilon \mid m \mid) \frac{Q_r(m_j)}{T(m_j)^s}$$

exists, for $y \neq 0$ on T_k, and is an entire function in s.

(ii) *The Dirichlet series*

$$\sum_{(m)}{}' \frac{Q_r(m_j)}{T(m_j)^s} = \zeta(s)$$

and the associated integral

$$\int_{|x| \geq 1} \frac{Q_r(x_j)}{T(x_j)^s}\, dv_x = H(s), \tag{2.9.5}$$

being both absolutely convergent in the right half-plane

$$\operatorname{Re} s > \frac{r+k}{2h}, \tag{2.9.6}$$

the holomorphic functions defined by them in this half-plane are such that their difference has an analytic continuation into the entire plane, that is,

$$\zeta(s) - H(s) = \text{entire function.}$$

(iii) *Taken separately, the functions $\zeta(s)$, $H(s)$ have meromorphic continuations into the entire s-plane. If $T(m_j)$ is homogeneous, then $H(s)$ and $\zeta(s)$ have (at most) one simple pole at the obvious point*

$$s = \frac{r+k}{2h}, \tag{2.9.7}$$

the residue being

$$\frac{1}{2h} \int_{S_k} \frac{Q_r(\xi_1, \ldots, \xi_k)\, d\omega}{T_{2h}(\xi_1, \ldots, \xi_k)^{r+k/2h}}. \tag{2.9.8}$$

If $T(m_j)$ is nonhomogeneous, there is first of all again a simple pole at the same point (2.9.7) *with 'the same' residue* (2.9.8), *where T_{2h} is the highest part of T; and there may also appear other poles, all simple ones, at the points*

$$s = \frac{r+k-m}{2h}, \quad m = 1, 2, 3, 4, \ldots, \tag{2.9.9}$$

which are placed at equal distances to the left of the first one.

Proof. Ad (iii). In the homogeneous case we have in the half-plane (2.9.6)

$$H(s)=\int_{t=1}^{\infty}\frac{t^{r+k-1}}{t^{2hs}}\,dt\,.\int_{S_k}\frac{Q_r(\xi)}{T_{2h}(\xi)^s}\,d\omega_\xi,$$

and in this product the first factor is $(2hs-r-k)^{-1}$ and the second is an entire function, which gives the conclusion.

If T is not homogeneous, we put

$$T=T_{2h}+P_{2h-1}=T_{2h}\left(1+\frac{P_{2h-1}}{T_{2h}}\right),$$

and if we substitute this in the integral (2.9.5) we obtain, at first formally,

$$H(s)=\int_{|x|\geqq 1}\frac{Q_r(x_j)}{T_{2h}^s}\sum_{n=0}^{\infty}\binom{-s}{n}\frac{P_{2h-1}^n}{T_{2h}^n}\,dv_x$$
$$=\sum_{n=0}^{\infty}\binom{-s}{n}\int_{|x|\geqq 1}\frac{Q_rP_{2h-1}^n}{T_{2h}^{s+n}}\,dv_x.$$

Now, the nth term in this sum is of the kind just discussed, with s replaced by $s+n$, and Q_r replaced by $Q_rP_{2h-1}^n$, and since the last expression is an inhomogeneous polynomial of degree $\leqq(2h-1)\,n$, we see that for $n>0$ the term has at most simple poles at the points

$$s+n=\frac{r+k+\nu}{2h},\quad 0\leqq\nu\leqq(2h-1)\,n,$$

and is holomorphic otherwise. Finally, corresponding to any bounded domain D of the s-plane we can find an M such that on putting

$$\left(1+\frac{P_{2h-1}}{T_{2h}}\right)^{-s}=\sum_{n=0}^{M}\binom{-s}{n}\left(\frac{P_{2h-1}}{T_{2h}}\right)^n+R_M(x_j;\,s)$$

the remainder $\displaystyle\int_{|x|\geqq 1}\frac{Q_r(x_j)}{T_{2h}^s}R_M(x_j;\,s)\,dv_x$

will be absolutely uniformly convergent for s in D and thus will be holomorphic. From all this our assertion follows.

Finally, returning to integrals instead of series, we take a function $\phi(\xi_j)$ on S_k as in theorem 2.7.6, say in $C^{(\infty)}$, and we put

$$f(x_j;\,s)=|\,x\,|^{a+bs}\,\phi(\xi_j)$$

with any real numbers a, $b\neq 0$, and we consider this in the half-plane $a+b.\mathrm{Re}\,s>-k$ for which the transform

$$\int_{Ek}e^{-2\pi i(x,\,y)}\,e^{-\epsilon|x|}f(x_j;\,s)\,dv_x$$

is definable. With our previous function $\chi(t)$ we now put

$$f(x_j;\, s) = \chi(|\, x\,|)f(x_j;\, s) + (1 - \chi(|\, x\,|)f(x_j;\, s) = f^1(x_j;\, s) + f^2(x_j;\, s),$$

and then put

$$g^r(y_j;\, s) = \lim_{\epsilon \downarrow 0} \int_{Ek} e^{-2\pi i(x,\, y)}\, e^{-\epsilon |x|} f^r(x_j;\, s)\, dv_x \tag{2.9.10}$$

for $r = 1, 2$. Now, $g^1(y_j;\, s)$ falls under theorem 2.7.6, and is therefore $C^{(\infty)}$ in y and holomorphic in s. For the function $g^2(y_j;\, s)$ this is, however, likewise so, simply because for this function the integration in (2.9.10) extends effectively only over the unit sphere $|\, x\,| \leqq 1$. Therefore, the function $g^1 + g^2$ is likewise so, and this proves a statement made about the function (2.7.9) near the end of section 2.7.

CHAPTER 3

CLOSURE PROPERTIES OF FOURIER TRANSFORMS

We will now present an analysis of the principal closure properties of characteristic functions in general, and for those of (infinitely) subdivisible processes in particular, and doing the latter immediately in E_k for general k will require a rather elaborate setting indeed; and such simple functions as are $e^{i(\alpha, x)}$ and $1-\cos(\alpha x)$ will be replaced by more general ones (which we will term 'pseudo-characters' and 'Poisson characters' respectively) for the sake of greater clarity, we hope.

3.1. Pseudo-characters and Poisson characters

DEFINITION 3.1.1. *A function*

$$\chi(\alpha; x) \equiv \chi(\alpha_1, \ldots, \alpha_k; x_1, \ldots, x_k) \tag{3.1.1.}$$

is a *pseudo-character* if (i) it is defined and continuous in $E_{2k} = E_k^\alpha \times E_k^x$ and is bounded there,

$$|\chi(\alpha; x)| \leqq M_0, \tag{3.1.2}$$

(ii) for $(\alpha) = (0)$, $\chi(\alpha; x)$ is a constant in x, but not $\equiv 0$,

$$\chi(0; x_j) = c_0 \neq 0, \tag{3.1.3}$$

and, what is decisive, (iii) we have

$$\lim_{|x| \to \infty} \int_{E_k} K(\alpha_j)\, \chi(\alpha_j; x_j)\, dv_\alpha = 0 \tag{3.1.4}$$

for every $K \in L_1(E_k)$.

THEOREM 3.1.1. *The class of pseudo-characters is closed under uniform convergence in E_{2k}, except for property* (ii).

In fact, if (3.1.4) holds for a sequence χ_n having a uniform limit χ_0, then it also holds for χ_0.

On the other hand, for fixed bounded χ, if (3.1.4) holds for a sequence K_n, for which $\| K_n - K \| \to 0$, then it also holds for K. Now, step functions on the finite intervals I_{ab}: $a_j \leqq x_j < b_j$ are dense in $L_1(E_k)$, and hence the following statement:

THEOREM 3.1.2. *A bounded continuous function $\chi(\alpha; x)$ fulfils* (3.1.4) *if we have*

$$\lim_{|x| \to \infty} \int_{I_{ab}} \chi(\alpha; x)\, dv_\alpha = 0 \tag{3.1.5}$$

for all finite multi-intervals.

In particular, we have

$$\int_{I_{ab}} e^{i(\alpha, x)}\, dv_\alpha = \prod_{j=1}^{k} \frac{e^{ib_j x_j} - e^{ia_j x_j}}{ix_j},$$

and it follows from $\left|\dfrac{e^{ibx} - e^{iax}}{ix}\right| \leqq \dfrac{2\,|b-a|}{1+|x|}$

that (3.1.5) holds, whence the following statement:

THEOREM 3.1.3. *The function* $e^{i(\alpha, x)}$ *is a pseudo-character* (*Riemann–Lebesgue lemma*).

Since $e^{i\lambda(\alpha, x)}$ is also a pseudo-character, for any λ, we may also state as follows:

THEOREM 3.1.4. *If* $B(t)$ *in* $-\infty < t < \infty$ *with* $B(0) \neq 0$, *is a sum*

$$\sum_{(m)} c_m e^{i\lambda_m t} \tag{3.1.6}$$

with

$$\lambda_m \neq 0, \quad \sum_{(m)} |c_m| < \infty,$$

or a uniform limit of such sums (*Bohr almost periodic function with* $\lambda_m \neq 0$), *then the function*

$$\chi(\alpha; x) \equiv B(\alpha_1 x_1 + \ldots + \alpha_k x_k) \tag{3.1.7}$$

is a pseudo-character.

For instance, $2\cos(\alpha, x) = e^{i(\alpha, x)} + e^{-i(\alpha, x)}$ is a pseudo-character.

DEFINITION 3.1.2. A *Poisson character* is a function $Q(\alpha; x)$ in $E_k \times E_k$ of the following description:

(i)

$$Q(\alpha; x) = c_1(1 - \chi(\alpha; x)), \quad c_1 > 0, \tag{3.1.8}$$

where $\chi(\alpha; x)$ is a pseudo-character, and $\chi(\alpha; x)$ is real valued, and

$$Q(\alpha; x) \geqq 0 \tag{3.1.9}$$

and also $Q(0; x) \equiv 0$ for all x, and $Q(\alpha; 0) \equiv 0$ for all α. (3.1.10)

(ii) The function

$$q(x) = \int_{|\beta| \leqq 1} Q(\beta; x)\, dv_\beta \tag{3.1.11}$$

is strictly >0 for $x \neq 0$ (and not only $\geqq 0$ as (3.1.9) implies) and

(iii) the quotient

$$P(\alpha; x) \equiv \frac{Q(\alpha; x)}{q(x)}, \quad x \neq 0, \tag{3.1.12}$$

is bounded and uniformly continuous in the variable α in the point set

$$0 \leqq |\alpha| < \alpha_0, \quad 0 < |x| \leqq 1 \tag{3.1.13}$$

for every $\alpha_0 > 0$.

Important remark. We do not assume that (3.1.12) is uniformly continuous in (3.1.13) as a function in (α, x) but only that it is uniformly continuous in α, uniformly in x. For $k=1$, for the classical function $Q(\alpha; x)=(\sin \alpha x)^2$ and for similar ones the quotient does happen to be continuous in (α, x) simultaneously, and all proofs, known to this author, of the structure theorem for subdivisible processes make full use of this simplification; but for $k \geqq 2$ this definitely ceases to be so and a more systematic approach similar to ours must be sought.

We note that in (3.1.11) and (3.1.13) the unit spheres could be replaced by others with fixed radii, equal or not, and that, on the other hand, the constant c_1 in (3.1.8) could be put equal to 1, which in the reasoning we will frequently do anyway.

Next, if $\omega_S(\beta)$ is the 'indicator function' (characteristic set function) of the unit sphere $|\beta| \leqq 1$, and if, on putting $c_1=1$, we write

$$q(x)=\int_{E_k} \omega_S(\beta)\,[1-\chi(\beta; x)]\,d\omega_\beta,$$

then we obtain

$$q(x)=c_2-p(x), \tag{3.1.14}$$

where $c_2>0$, and the function

$$p(x)=\int_{E_k} \omega_S(\beta)\,\chi(\beta; x)\,d\omega_\beta$$

tends to 0 as $|x| \to \infty$ by definition 3.1.1. Therefore, in connection with property (ii) of definition 3.1.2 we obtain the following further property automatically:

(iv) For any fixed $\eta>0$ we have

$$M_1<q(x)<M_2 \tag{3.1.15}$$

for $|x| \geqq \eta$, with $0<M_1<M_2<\infty$.

THEOREM 3.1.5. *If $B(t)$ in $-\infty<t<\infty$ is periodic or Bohr almost periodic, if it is non-negative and even,*

$$B(t) \geqq 0, \quad B(-t)=B(t), \tag{3.1.16}$$

and if in a neighborhood of $t=0$ we have

$$B(t)=c_2\,|t|^m+o(|t|^m) \tag{3.1.17}$$

with $c_2>0$, $m>0$ (m not necessarily integer), then

$$Q(\alpha; x)=B(\alpha_1 x_1+\ldots+\alpha_k x_k) \tag{3.1.18}$$

is a Poisson character.

Also we then have $$q(x)=c_3\,|\,x\,|^m+o(|\,x\,|^m) \tag{3.1.19}$$
for $|\,x\,|\to 0$ *and*

$$P(\alpha;\,x)=\frac{c_2}{c_3}\left|\alpha_1\frac{x_1}{|\,x\,|}+\ldots+\alpha_k\frac{x_k}{|\,x\,|}\right|^m+P^*(\alpha;\,x), \tag{3.1.20}$$

where $P^*(\alpha;\,x)$ *is continuous in* (α, x) *for* $\alpha\in E_k$, $x\neq 0$ *and*

$$\lim_{|\,x\,|\to 0} P^*(\alpha;\,x)=0 \tag{3.1.21}$$

uniformly in $|\,\alpha\,|\leqq\alpha_0$ *for any* $\alpha_0>0$.

Proof. For an almost periodic $B(t)$ there exists the mean value

$$c=\lim_{T\to\infty}\frac{1}{T}\int_0^T B(t)\,dt$$

which in our case must be >0, and if we put $B(t)=c(1-B_0(t))$ then $\chi(\alpha;\,x)\equiv B_0(\alpha_1x_1+\ldots+\alpha_kx_k)$ is a pseudo-character, first of all.

Next, if in the β-space we introduce polar coordinates (spherical), and if for given vectors (x_j) and (β_j) we denote the angle between them by θ, then we have

$$\int_{|\beta|\geqq 1}|\,(\beta, x)\,|^m\,dv_\beta=\int_{|\beta|\geqq 1}|\,\beta\,|^m.\,|\,x\,|^m.\,(\cos\theta)^m\,dv_\beta$$
$$=|\,x\,|^m\int_{|\beta|\geqq 1}|\,\beta\,|^m.\,|\cos\theta\,|^m\,dv_\beta=c_5\,|\,x\,|^m,$$

and therefore also, if $(x)\to 0$,

$$\int_{|\beta|\geqq 1}|\,o\,(|(\beta, x)\,|)\,|^m\,dv_\beta=o(|\,x\,|^m),$$

which proves (3.1.19). Finally, an easy discussion of $\dfrac{o(|(\alpha, x)\,|^m)}{|\,x\,|^m}$ will prove (3.1.21).

3.2. Pseudo-transforms and positive definite functions

We take the set of all continuous complex-valued functions in E_k: (α_j) and we define as follows:

Definition 3.2.1. A sequence $\{\psi_n(\alpha)\}$ is called *P-convergent* (towards its P-limit), in symbols

$$\psi_n(\alpha)\overset{P}{\to}\psi(\alpha), \tag{3.2.1}$$

if we have $$\lim_{n\to\infty}\psi_n(\alpha)=\psi(\alpha) \tag{3.2.2}$$

uniformly in every compact sphere

$$0\leqq|\,\alpha\,|\leqq\alpha_0,\quad \alpha_0>0. \tag{3.2.3}$$

A set $\{\psi(\alpha)\}$ will be called P-closed if it contains the P-limits of its sequences. The set of all continuous functions is P-closed and therefore to any subset $S=\{\psi(\alpha)\}$ there corresponds a smallest P-closed set $\tilde{S}$ containing it ('the P-closure of S'), and it is easily seen that $\tilde{S}$ arises by adding to S all the P-limits of its sequences.

DEFINITION 3.2.2. Given a *fixed* pseudo-character, a function $\phi(\alpha_j)$ will be called a *pseudo-transform* if it can be represented with some F in V^+ in the form

$$\phi \equiv \phi^F(\alpha) = \int_{E_k} \chi(\alpha; x)\, dF(x). \tag{3.2.4}$$

Obviously, $\phi(\alpha)$ is bounded,

$$|\phi(\alpha)| \leqq M_0 . \|F\|, \tag{3.2.5}$$

and is continuous in α although perhaps not uniformly continuous in E_k.

If we are given a sequence

$$\phi_n(\alpha) = \int_{E_k} \chi(\alpha; x)\, dF_n(x), \tag{3.2.6}$$

then
$$\phi_n(0) = c_0 F_n(E_k) \tag{3.2.7}$$

implies that we have
$$|F_n(E_k)| \leqq M_1 \tag{3.2.8}$$

whenever
$$|\phi_n(0)| \leqq M; \tag{3.2.9}$$

and also if a sequence (3.2.6) converges a.e. and (3.2.8) holds, then the sequence converges boundedly a.e.

If the sequence F_n is weakly convergent, then the sequence $\phi_n(\alpha)$ need not converge at all, but if F_n is Bernoulli convergent to F_0, then $\phi_n(\alpha)$ is P-convergent to

$$\phi_0(\alpha) = \int_{E_k} \chi(\alpha; x)\, dF_0(x) \tag{3.2.10}$$

(see lemma 1.5.1).

THEOREM 3.2.1. (i) *The set of pseudo-transforms is P-closed. Also, if a sequence* (3.2.6) *is convergent at all points,*

$$\phi_n(\alpha) \to \phi(\alpha), \tag{3.2.11}$$

and if $\phi(\alpha)$ is continuous at the origin, then for a subsequence $\{n_k\}$

$$\phi_{n_k}(\alpha) \overset{P}{\to} \phi(\alpha) \tag{3.2.12}$$

and F_{n_k} is Bernoulli convergent.

(ii) *If* (3.2.11) *holds a.e. and also* (3.2.9) *holds, then F_n converges weakly to F_0 and $\phi(\alpha) = \phi_0(\alpha)$ a.e.*

Proof. Due to (3.2.9), and hence (3.2.8), the sequence F_n has a weakly convergent subsequence. We denote the latter by F'_n and its limit by F'_0, and we introduce the transforms

$$\phi'_n(\alpha)=\int \chi(\alpha;x)\,dF'_n(x), \quad n=0,1,2,3,\ldots. \tag{3.2.13}$$

With a family of kernels $\{K_R(\alpha_j)\}$ in L_1 we introduce the functions

$$\phi'_{nR}(\alpha)=\int K_R(\alpha-\beta)\,\phi'_n(\beta)\,dv_\beta, \quad \chi_R(\alpha;x)=\int K_R(\alpha-\beta)\,\chi(\beta;x)\,dv_\beta,$$

which are connected by

$$\phi'_{nR}(\alpha)=\int \chi_R(\alpha;x)\,dF'_n(x).$$

Now, by the very definition of a pseudo-character, $\chi_R(\alpha;x)\to 0$ as $|x|\to\infty$, uniformly in (3.2.3), and therefore

$$\phi'_{nR}(\alpha)\overset{P}{\to}\phi'_{0R}(\alpha).$$

However, if (3.2.11) holds, then

$$\lim_{n\to\infty}\phi'_{nR}(\alpha)=\phi_R(\alpha)\equiv\int K_R(\alpha-\beta)\,\phi(\beta)\,dv_\beta,$$

and thus we have $$\phi'_{0R}(\alpha)=\phi_R(\alpha).$$

Letting $R\to\infty$, we hence obtain by theorem 1·1

$$\phi'_0(\alpha)=\phi(\alpha)$$

at all points where both functions are continuous. However, $\phi'_0(\alpha)$ is continuous by construction and $\phi(\alpha)$ is continuous at $\alpha=0$ by assumption, so that we have $\phi'_0(0)=\lim_{n\to x}\phi'_n(0)$. By (3.2.7) we thus have $F'_0(E_k)=\lim_{n\to\infty}F'_n(E_k)$, and thus F'_n is even Bernoulli convergent and therefore $$\phi'_n(\alpha)\overset{P}{\to}\phi'_0(\alpha)\equiv\phi(\alpha).$$

Now, by explicit assumption, the entire sequence $\phi_n(\alpha)$ is convergent to $\phi(\alpha)$, and by what we have just proven, any subsequence of $\phi_n(\alpha)$ must contain a sub-subsequence which is P-convergent and the corresponding F_{n_k} are Bernoulli convergent.

For the second part of the theorem we take any continuous function $C(\beta)$ which is zero outside a compact set and obtain

$$\int C(\beta)\,\phi'_0(\beta)\,dv_\beta=\int C(\beta)\,\phi(\beta)\,dv_\beta,$$

and thus $\phi'_0(\beta)=\phi(\beta)$, a.e. as claimed.

Theorem 3.2.1 applies in particular to transforms proper

$$\phi(\alpha) \equiv \phi^F(\alpha) = \int_{E_k} e^{-2\pi i(x,\alpha)} dF(x), \tag{3.2.14}$$

and for these the following property is supplementary:

THEOREM 3.2.2. *For $F \in V^+$ we have*

$$|\phi^F(\alpha+h) - \phi^F(\alpha)|^2 \leqq 2 \|F\| \cdot \text{ real part } (\phi^F(0) - \phi^F(h)), \tag{3.2.15}$$

and, more generally, for $F \in V$, if $\tilde{F}$ is the absolute value of F and $\tilde{\phi}$ is its transform then we have

$$|\phi(\alpha+h) - \phi(\alpha)|^2 \leqq 2 \|F\| \cdot \text{ real part } (\tilde{\phi}(0) - \tilde{\phi}(h)). \tag{3.2.16}$$

Proof.

$$|\phi(\alpha+h) - \phi(\alpha)|^2 \leqq \left(\int_{E_k} |e^{-2\pi i(h,x)} - 1| \, d\tilde{F}\right)^2 = \left(2\int_{E_k} |\sin \pi(h,x)| \, d\tilde{F}\right)^2$$

and by Schwarz's inequality this is

$$\leqq 4\int_{E_k} (\sin \pi(h,x))^2 d\tilde{F} \,.\int_{E_k} d\tilde{F} = 2 \|\tilde{F}\| \int_{E_k} (1 - \cos 2\pi(h,x)) \, d\tilde{F},$$

as claimed.

THEOREM 3.2.3. *In order that a continuous function $\phi(\alpha_j)$ be presentable in the form (3.2.14), with $F \in V^+$, it is necessary and sufficient that it be positive definite in the sense that we have*

$$\sum_{p,q=1}^{N} \phi(\alpha^p - \alpha^q) \rho_p \overline{\rho_q} \geqq 0, \tag{3.2.17}$$

for any finitely many points $\alpha^1, \alpha^2, \ldots, \alpha^N$ in E_k, and any complex numbers $\rho_1, \ldots, \rho_N$.

Proof. For an integral (3.2.14) the left side in (3.2.17) has the value

$$\int \left| \sum_{p=1}^{N} \rho_p e^{-2\pi i(\alpha^p, x)} \right|^2 dF(x),$$

which is indeed $\geqq 0$.

Conversely, if we start from (3.2.17), then using it for two points $(\alpha, 0)$ and arbitrary complex numbers ρ, σ, it follows that $|\phi(\alpha)| \leqq \phi(0)$, so that $\phi(\alpha)$ is bounded, and this being so, the 'discrete' condition (3.2.17) implies its 'integrated' counterpart

$$\int_{E_k} \int_{E_k} \phi(\alpha, \beta) \rho(\alpha) \rho(\overline{\beta}) \, dv_\alpha dv_\beta \geqq 0 \tag{3.2.18}$$

for any $\rho(\alpha) \in L_1(E_k)$. For fixed x and $\epsilon > 0$ we put

$$\rho(\alpha) = e^{-2\epsilon|\alpha|^2} e^{2\pi i(\alpha, x)},$$

so that

$$\iint e^{-2\epsilon(|\alpha|^2+|\beta|^2)} \phi(\alpha-\beta) e^{2\pi i(\alpha-\beta, x)} dv_\alpha dv_\beta \geqq 0,$$

and if we make the change of variables (in vectors)

$$\alpha - \beta = \gamma, \quad \alpha + \beta = \delta,$$

then this implies

$$\int_{E_k} \left(\int_{E_k} e^{-\epsilon|\delta|^2} dv_\delta \right) e^{-\epsilon|\gamma|^2} \phi(\gamma) e^{2\pi i(\gamma, x)} dv_\gamma \geqq 0.$$

This means that the function $\phi_\epsilon(\gamma) = e^{-\epsilon|\gamma|^2}\phi(\gamma)$, which for $\epsilon > 0$ belongs to L_1, has a non-negative (anti)-transform, and theorem 2.2.1 implies that $\phi_\epsilon(\alpha)$ has a representation (3.2.14). Letting $\epsilon \to 0$, this then also applies to $\phi(\alpha)$ itself by theorem 3.2.1.

The proof just completed also implies the following statement:

THEOREM 3.2.4. *If a bounded measurable function $\phi(\alpha)$ satisfies condition* (3.2.18), *then it differs from a continuous positive definite function on a set of measure zero.*

3.3. Poisson transforms

DEFINITION 3.3.1. Given a fixed Poisson character $Q(\alpha; x)$ we call a function $\psi(\alpha_j)$ a *Poisson transform* if it is continuous and can be represented by an integral

$$\psi(\alpha_j) = \int_{E_k'} Q(\alpha_j; x_j)\, dF(x_j), \quad F(A) \geqq 0, \tag{3.3.1}$$

where E_k' is the set $0 < |x| < \infty$ in E_k, that is, the set which arises from E_k by deletion of the origin. (For a description of the integral see section 4.1.)

THEOREM 3.3.1. (i) *If an integral* (3.3.1) *is finite on an α-set of positive measure then we have*

$$\int_{|x| \geqq r_0} dF(x) < \infty \tag{3.3.2}$$

for some, and hence every, $r_0 > 0$.

(ii) *If the integral* (3.3.1) *is bounded in* $|\beta| \leqq 1$, *or only if*

$$\int_{|\beta| \leqq 1} \psi(\beta)\, d(\beta) < \infty, \tag{3.3.3}$$

then

$$\int_{E_k'} q(x_j)\, dF(x) < \infty, \tag{3.3.4}$$

or what is equivalent with it,

$$\int_{S_k'} q(x)\,dF(x) + \int_{|x|>1} dF(x) < \infty, \tag{3.3.5}$$

when $S_k' = \{0 < |x| \leqq 1\}$.

(iii) *Conversely, if* (3.3.5) *holds then* $\psi(\alpha)$ *is finite and continuous and thus a Poisson transform.*

Proof. If (3.3.1) is finite on a set of positive measure then there is an $N > 0$ and a Borel set A with $0 < v(A) < \infty$ such that $\psi(\alpha) \leqq N$ for $\alpha \in A$. Now, by definition 3.1.1 the function

$$\chi(x) \equiv \int_{E_k} \omega_A(\beta)\,\chi(\beta; x)\,dv_\beta = \int_A \chi(\beta; x)\,dv_\beta$$

tends to 0 as $|x| \to \infty$, and thus we have

$$Nv(A) \geqq \int_A \psi(\beta)\,dv_\beta = \int_{E_k'} \left(\int_A Q(\beta; x)\,dv_\beta \right) dF(x) = \int_{E_k'} [v(A) - \chi(x)]\,dv_x$$

$$\geqq \int_{|x| \geqq r_0} (v(A) - \chi(x))\,dv_x \geqq \tfrac{1}{2} v(A) \int_{|x| \geqq r_0} dF(x),$$

for r_0 sufficiently large, which proves part (i). Part (ii) follows obviously from

$$\int_{|\beta| \leqq 1} \psi(\beta)\,dv_\beta = \int_{E_k'} q(x)\,dF(x) \tag{3.3.6}$$

by part (ii) of definition 3.1.2. Finally, if in E_k' we introduce the set function $G(A) = \int_A q(x)\,dF(x)$, then (3.3.4) means that $G(E_k') < \infty$, and, on the other hand, we can write $\psi(\beta) = \int_{E_k'} P(\beta; x)\,dG(x)$, and part (iii) follows from the properties stipulated for $P(\beta; x)$ in definition 3.1.2.

We are now going to make statements on (3.3.1) in case $F(A)$ is zero either in $0 < |x| < 1$ or in $|x| > 1$.

THEOREM 3.3.2. *The Poisson transforms of the form*

$$\psi^2(\alpha,) = \int_{|x| \geqq 1} Q(\alpha; x)\,dF(x) \tag{3.3.7}$$

are a P-closed set.

Proof. If a sequence

$$\psi_n^2(\alpha) = \int_{|x| \geqq 1} Q(\alpha; x)\,dF_n(x) \tag{3.3.8}$$

is P-convergent towards a function $\psi(\alpha)$, then by (3.3.6) we have $F_n(E_k) \leqq M$, and since it suffices to show that the limit of a subsequence of (3.3.8) is of the form (3.3.7) we may immediately add the assumption that F_n is weakly convergent towards some $F(A)$ and that the numbers $\gamma_n = F_n(E_k)$ are convergent towards some number γ. Thus the functions

$$\gamma_n - \psi_n^2(\alpha) = \int_{|x| \geqq 1} \chi(\alpha; x)\, dF_n(x)$$

are a P-convergent sequence, and by theorem 3.2.1 we therefore have

$$\gamma - \psi^2(\alpha) = \int_{|x| \geqq 1} \chi(\alpha; x)\, dF(x),$$

and $F_n(E_k) \to F(E_k) \equiv \gamma$, which proves the theorem.

DEFINITION 3.3.2. A set $\{\psi(\alpha)\}$ will be called P-finite if every sequence $\{\psi_n(\alpha)\}$ in it for which

$$\sup_{|\beta| \leqq 1} |\psi_n(\beta)| \leqq M_1 \tag{3.3.9}$$

is equi-uniformly continuous in $|\alpha| \leqq \alpha_0$ for every $\alpha_0 > 0$.

Now, equality (3.3.6) and our assumptions on $P(\alpha; x)$ easily imply as follows:

THEOREM 3.3.3. *The set of Poisson transforms of the form*

$$\psi^1(\alpha) = \int_{S_k'} Q(\alpha; x)\, dF(x) \tag{3.3.10}$$

is P-finite, and for a sequence

$$\psi_n^1(\alpha) = \int_{S_k'} Q(\alpha; x)\, dF_n(x) \tag{3.3.11}$$

the relation (3.3.9) *is equivalent with*

$$\int_{S_k'} q(x)\, dF_n(x) \leqq M_2. \tag{3.3.12}$$

DEFINITION 3.3.3. We call a function in E_k: (α_j) *pseudo-Gaussian*, and we denote it by $R(\alpha)$, if it is a P-limit of a sequence of Poisson transforms

$$\psi_n^0(\alpha) = \int_{0 < |x| \leqq r_n} Q(\alpha; x)\, dF_n(x) \tag{3.3.13}$$

with $r_n \to 0$.

Any function (3.3.13) is a special case of (3.3.10), and it is not hard to establish the following theorem:

THEOREM 3.3.4. *The set of pseudo-Gaussian functions* $\{R(\alpha)\}$ *is P-finite and P-closed.*

But now we are coming to a key theorem.

THEOREM 3.3.5. *The set of functions*

$$R(\alpha)+\psi^1(\alpha), \tag{3.3.14}$$

where $R(\alpha)$ *is pseudo-Gaussian and* $\psi^1(\alpha)$ *is of the form* (3.3.10) *is the* (*smallest*) P-*closure of the set of functions* $\{\psi^1(\alpha)\}$.

Proof. It is quite easy to argue that the P-closure of (3.3.10) must include all functions (3.3.14), but we must show conversely that if a sequence (3.3.11) is P-convergent towards a function $\psi_0^1(\alpha)$, then there exists a function (3.3.10) and a function $R(\alpha)$ such that

$$\psi_0^1(\alpha)=R(\alpha)+\psi^1(\alpha). \tag{3.3.15}$$

Now, the P-convergence of (3.3.11) implies (3.3.12), and, after passing to a subsequence if necessary, we may assume that there is a weak limit $F(x)$ in the sense that, again, $\int_{S_k'} q(x)\,dF(x) \leqq M_2$ and

$$\int_{S_k'} c(x)\,dF_n(x) \to \int_{S_k'} c(x)\,dF(x), \tag{3.3.16}$$

for every continuous $c(x)$ in S_k' which vanishes in some neighborhood of the origin. Now, the function in r,

$$\int_{r\leqq x\leqq 1} dF(x), \quad 0<r<1,$$

is monotonely increasing as r decreases. Therefore, it is continuous for all but countably many values r, and if we keep those excluded then (3.3.16) implies

$$\int_{r\leqq x\leqq 1} Q(\alpha;\,x)\,dF_n(x) \overset{P}{\to} \int_{r\leqq x\leqq 1} Q(\alpha;\,x)\,dF(x). \tag{3.3.17}$$

We now form $\psi^1(\alpha)$ with this $F(x)$, and our theorem will be proven if we show that the difference

$$R(\alpha)=\psi_0^1(\alpha)-\psi^1(\alpha)$$

is pseudo-Gaussian.

For each $p=1,2,\ldots$, we can pick a value $r_p<1/p$ such that

$$\int_{0<|x|\leqq r_p} Q(\alpha;\,x)\,dF(x)\leqq\frac{1}{p} \quad \text{for } |\alpha|\leqq p. \tag{3.3.18}$$

But, for fixed r_p we can find an index n_p such that

$$|\psi_0^1(\alpha)-\psi_{n_p}^1(\alpha)|+\left|\int_{r_p<|x|\leqq 1} Q(\alpha,x)\,d(F_{n_p}-F)\right|\leqq\frac{1}{p} \tag{3.3.19}$$

for $|\alpha| \leqq p$, and from (3.3.18) and (3.3.19) we can put together the estimate

$$\left| R(\alpha) - \int_{0<|x|\leqq r_p} Q(\alpha; x)\, dF_{n_p} \right| \leqq \frac{2}{p}$$

for $|\alpha| \leqq p$, which shows indeed that $R(\alpha)$ is a P-limit of a sequence (3.3.13).

THEOREM 3.3.6. *For the set of Poisson transforms $\{\psi(\alpha)\}$ the P-closure is the set of all functions*

$$\{R(\alpha) + \psi(\alpha)\}. \tag{3.3.20}$$

Proof. Obviously all functions (3.3.20) are in the P-closure. Take now conversely a P-convergent sequence of Poisson transforms $\psi_n(\alpha)$ and put $\psi_n(\alpha) = \psi_n^1(\alpha) + \psi_n^2(\alpha)$, where

$$\psi_n^1(\alpha) = \int_{S_k'} Q(\alpha, x)\, dF_n(x), \quad \psi_n^2(\alpha) = \int_{|x|>1} Q(\alpha; x)\, dF_n(x).$$

By what we already know, after passing to a subsequence, we may assume that the sequence $\{\psi_n^1(\alpha)\}$ is P-convergent, its P-limit being of the form

$$R(\alpha) + \int_{S_k'} Q(\alpha; x)\, dF^1(x). \tag{3.3.21}$$

The differences $\psi_n^2(\alpha) = \psi_n(\alpha) - \psi_n^1(\alpha)$ are therefore also P-convergent, the limit being of the form

$$\int_{|x|\geqq 1} Q(\alpha; x)\, dF^2(x). \tag{3.3.22}$$

But the sum of functions (3.3.21) and (3.3.22) is of the form (3.3.20), q.e.d.

THEOREM 3.3.7. (i). *If $Q(\alpha; x)$ is as in theorem* 3.1.5,

$$Q(\alpha; x) \equiv B(\alpha_1 x_1 + \ldots + \alpha_k x_k), \tag{3.3.23}$$

then the set of pseudo-Gaussian functions $\{R(\alpha)\}$ is the P-closure of the finite linear combinations with positive coefficients of the 'monomials',

$$|\alpha_1 \xi_1 + \ldots + \alpha_k \xi_k|^m \tag{3.3.24}$$

with $\xi_1^2 + \ldots + \xi_k^2 = 1$, say.

(ii) *For $m=2$, $\{R(\alpha)\}$ are all non-negative quadratic forms*

$$\sum_{p,q=1}^{k} c_{pq} \alpha_p \alpha_q \geqq 0.$$

(iii) *In the representation of a closure element $\psi_0(\alpha)$ as a sum $R(\alpha) + \psi(\alpha)$ the two addends $R(\alpha)$, $\psi(\alpha)$ are uniquely determined.*

Proof. Ad (i). In general, if we can put $Q(\alpha; x) = Q^1(\alpha; x) + Q^2(\alpha; x)$, where

$$\lim_{|x|\to 0} \frac{Q^2(\alpha; x)}{q(x)} = 0$$

uniformly in $|\alpha| \leqq \alpha_0$, for every $\alpha_0 > 0$, then in the construction of an $R(\alpha)$ as a limit of a sequence (3.3.13) we may replace each element of the sequence by

$$\int_{0<|x|\leqq r_n} Q^1(\alpha; x)\, dF_n(x),$$

without altering the fact that it is P-convergent, and towards $R(\alpha)$.

Now, in our present case this applies with $Q^1(\alpha; x) = |(\alpha, x)|^m$, but every integral

$$\int_{0<|x|\leqq r_n} |\alpha_1 x_1 + \ldots + \alpha_k x_k|^m\, dF_n(x), \tag{3.3.25}$$

is a P-limit of approximating Riemann expressions each of which can be written as a linear combination, with positive coefficients, of terms (3.3.24).

Conversely, the pseudo-Gaussian functions are a (semi)-vector field with positive coefficients, but each term (3.3.24) is a pseudo-Gaussian, since it is a P-limit of expressions (3.3.25), all having identical values, and corresponding to set functions $F_n(A)$ having the value $1/r_n^m$ at the point $r_n \xi_1, \ldots, r_n \xi_k$, and being zero everywhere else.

Ad (ii). For $m=2$ these functions are precisely all quadratic forms $\geqq 0$.

Ad (iii). If we envisage two representations

$$R_1(\alpha) + \psi_1(\alpha) = R_2(\alpha) + \psi_2(\alpha),$$

then on putting $R_1(\alpha) - R_2(\alpha) = \psi_2(\alpha) - \psi_1(\alpha)$ we obtain an identity

$$R(\alpha_1, \ldots, \alpha_k) = \int_{E_k'} B(\alpha_1 x_1 + \ldots + \alpha_k x_k)\, dF(x), \tag{3.3.26}$$

in which $R(\alpha_j)$ is a homogeneous function of weight m, and

$$F(A) = F_1(A) - F_2(A)$$

is a set function of mixed sign for which we have

$$\int_{S_k'} |x|^m\, |dF(x)| + \int_{|x|>1} |dF(x)| < \infty. \tag{3.3.27}$$

If we divide (3.3.26) by $|\alpha|^m$ we obtain

$$R\left(\frac{\alpha_1}{|\alpha|}, \ldots, \frac{\alpha_k}{|\alpha|}\right) \leqq \int_{0<|x|\leqq\delta} \frac{B((\alpha, x))}{|\alpha|^m . |x|^m} |x|^m\, |dF(x)|$$
$$+ \frac{1}{|\alpha|^m} \int_{|x|>\delta} B(\alpha, x)\, |dF(x)|.$$

Now, the quotient $B((\alpha, x))/|\alpha|^m |x|^m$ is bounded in (α, x), and therefore the first integral is $\leqq \epsilon$ for $\delta \leqq \delta(\epsilon)$ for all α. But, for fixed δ, since $B((\alpha, x))$ is bounded, the second integral tends to 0 as $|\alpha| \to \infty$, and therefore the left side tends to 0 as $|\alpha| \to \infty$. But the left side depends only on the ratios $\alpha_1 : \alpha_2 : \ldots : \alpha_k$, and must be identically 0 therefore.

3.4. Infinitely subdivisible processes

We take the same Poisson character $Q(\alpha; x)$ as before and the function $q(x)$ obtained from it, and in addition to that a real-valued function $\tilde{Q}(\alpha; x)$, not necessarily a character, which is continuous in (α, x), and bounded in x for $|\alpha| \leqq \alpha_0$, any $\alpha_0 > 0$, and for which we have

$$\lim_{|x| \to 0} \frac{\tilde{Q}(\alpha; x)}{q(x)} = 0, \tag{3.4.1}$$

uniformly in $|\alpha| \leqq \alpha_0$.

THEOREM 3.4.1. *The 'transform'*

$$\psi(\alpha) + i\tilde{\psi}(\alpha) = \int_{E'_k} (Q(\alpha; x) + i\tilde{Q}(\alpha; x))\, dF(x) \tag{3.4.2}$$

exists for

$$F(A) \geqq 0, \quad \int_{E'_k} q(x)\, dF(x) < \infty, \tag{3.4.3}$$

and for the set of functions (3.4.2) *the P-closure is the set of functions*

$$R(\alpha) + \psi(\alpha) + i\tilde{\psi}(\alpha), \tag{3.4.4}$$

where $\{R(\alpha)\}$ *are the same pseudo-Gaussian functions as before.*

Proof. In other words, the imaginary parts $\tilde{\psi}(\alpha)$ do not create pseudo-Gaussian functions of their own, and this, as we will see, is simply due to assumption (3.4.1), which, as can be readily seen, first of all secures the existence and continuity of the imaginary parts $\tilde{\psi}(\alpha)$ whenever $F(A)$ satisfies the requirement (3.4.3), although this requirement was arrived at to suit the needs of the real parts $\psi(\alpha)$, originally.

Now, if we are given a P-convergent sequence $\psi_n + i\tilde{\psi}_n$, then we again dissect it into the parts

$$\psi_n^1 + i\tilde{\psi}_n^1 = \int_{S'_k} (Q + i\tilde{Q})\, dF_n, \tag{3.4.5}$$

$$\psi_n^2 + i\tilde{\psi}_n^2 = \int_{|x|>1} (Q + i\tilde{Q})\, dF_n, \tag{3.4.6}$$

and the sequence (3.4.5) is P-finite again. Therefore, after passing to a subsequence, we may assume that in S'_k the sequence F_n is weakly convergent, and we now obtain

$$\psi_n^1 + i\tilde{\psi}_n^1 \overset{P}{\to} R(\alpha) + \int_{S'_k} (Q + i\tilde{Q})\, dF^1(A),$$

because formula (3.1.16) implies that we have

$$\int_{S'_k} \tilde{Q}\, dF_n \overset{P}{\to} \int_{S'_k} \tilde{Q}\, dF^1,$$

on account of (3.4.1). This being so, the sequence (3.4.6) is now likewise P-convergent, and so are therefore the real parts ψ_n^2. Therefore, the sequence $F_n(A)$ if envisaged on the closed set $|x| \geqq 1$ is Bernoulli convergent there, towards a function $F^2(A)$, and thus

$$\psi_n^2 + i\tilde{\psi}_n^2 \overset{P}{\to} \int_{|x| \geqq 1} (Q + i\tilde{Q})\, dF^2(A),$$

which completes the proof.

In the classical case we are given, to start with, the combination

$$Q + iQ' = 1 - e^{-2\pi i(\alpha, x)} = 2(\sin \pi(\alpha, x))^2 + i \sin 2\pi(\alpha, x)$$

in which, however, Q' does not satisfy requirement (3.4.1). But at the expense of adding a linear term in the α's outside the integral, this can be corrected by putting

$$\tilde{Q}(\alpha; x) = 2 \sin \pi(\alpha, x) - 2\pi(\alpha, \lambda(x)), \tag{3.4.7}$$

where in $(\alpha, \lambda) = \alpha_1 \lambda_1(x) + \ldots + \alpha_k \lambda_k(x)$ we may take for $\lambda_j(x)$ any continuous function in E_k, for which

$$\lambda_j(x) = \begin{cases} x_j + O(|x|^2) & \text{as} \quad |x| \to 0 \\ O(1) & \text{as} \quad |x| \to \infty \end{cases}$$

holds. In the theory of probability it has become customary to put $\lambda_j(x) = x_j(1 + |x|^2)^{-1}$, and in the theory of generalized Fourier integrals another normalization is being used but no choice has a stochastic preference over any other, and in order to emphasize this we will leave the functions $\lambda_j(x)$ unspecified, although we assume that they have been chosen fixed.

We are introducing the linear form

$$\psi^B(\alpha) = \sum_{p=1}^{k} c_p \alpha_p \tag{3.4.8}$$

for any real c_p (B for Bernoulli), the general real symmetric form

$$\psi^G(\alpha) = \sum_{p,q=1}^{k} c_{pq} \alpha_p \alpha_q \geqq 0, \tag{3.4.9}$$

(G for Gauss) and the general function

$$\psi^P(\alpha) = \int_{E'_k} (1 - e^{-2\pi i(\alpha, x)} - 2\pi i(\alpha, \lambda(x))\, dF(x) \qquad (3.4.10)$$

for any $\quad F(A) \geqq 0, \quad \int_{S'_k} |x|^2 dF(x) + \int_{|x|>1} dF(x) < \infty \qquad (3.4.11)$

(P for Poisson) and we claim as follows:

THEOREM 3.4.2. *If we start out with the set of functions*

$$\psi^0(\alpha) = \int_{E_k} (1 - e^{-2\pi i(\alpha, x)})\, dF(x) \qquad (3.4.12)$$

$$F(A) \geqq 0, \quad F(E_k) < \infty,$$

then its P-closure is the set of functions

$$\psi(\alpha) = i\,\psi^B(\alpha) + \psi^G(\alpha) + \psi^P(\alpha), \qquad (3.4.13)$$

with all Bernoulli, Gauss and Poisson addends used.

Also, we have $\quad \psi^P(\alpha) = o(|\alpha|^2) \quad$ as $\quad |\alpha| \to \infty, \qquad (3.4.14)$

and the quantities $c_p, c_{pq}, F(A)$ *in the decomposition* (3,4,13) *are uniquely determined.*

Proof. If a sequence

$$\psi_n^0(\alpha) = \int_{E_k} (1 - e^{-2\pi i(\alpha, x)})\, dF_n(x) \qquad (3.4.15)$$

is P-convergent then so are its real parts by themselves. Therefore

$$\int_{S'_k} |x|^2 dF_n + \int_{|x|>1} dF_n \leqq M,$$

and if we put each function (3.4.15) in the form

$$i \sum c_p \alpha_p + \int_{E'_k} (1 - e^{-2\pi i(\alpha, x)} - 2\pi i(\alpha, \lambda))\, dF(x), \qquad (3.4.16)$$

then it follows that the imaginary parts of the integral and the linear parts outside are both P-finite. Therefore, in a subsequence, the linear parts and the imaginary parts will be P-convergent, and by using our previous results it is not hard to see that the P-limits of (3.4.15) will be the functions (3.4.13) with all possible $\psi^G(\alpha)$, and $\psi^P(\alpha)$ and some $\psi^B(\alpha)$. But actually, we can obtain all possible linear parts $\psi^B(\alpha)$, since any term $i\alpha_j$ is the P-limit of $1/n(1 - e^{-i\alpha_j/n})$ as $n \to \infty$.

Next we note that for $|x| \leqq 1$, $|\alpha| \geqq 1$ we have

$$\frac{|Q+i\tilde{Q}|}{|\alpha|^2 . |x|^2} \leqq M$$

and (3.4.14) follows, by putting

$$\frac{\psi^P(\alpha)}{|\alpha|^2} = \int_{0<|x|\leqq\delta} \frac{Q+i\tilde{Q}}{|\alpha|^2 . |x|^2} |x|^2 dF(x) + \frac{1}{|\alpha|^2} \int_{|x|>\delta} (Q+iQ)\, dF(x)$$

and letting $|\alpha| \to \infty$ for δ small though fixed.

As regards uniqueness we note that if we consider the real part only, then the addend $\psi^G(\alpha)$ is uniquely determined by part (iii) of theorem 3.3.7, and we will have to show that if a function $\psi(\alpha)$ is of the form (3.4.16) and $F(A)$ satisfies the general assumption (3.4.11) then $F(A)$ is uniquely determined by this. With any vector

$$h = (h_1, \ldots, h_k)$$

we form

$$\begin{aligned} \frac{1}{4} \sum_{j=1}^{k} [\psi(\alpha_1, \ldots, \alpha_j + h_j, \ldots, a_k) - 2\psi(\alpha_1, \ldots, \alpha_j, \ldots, \alpha_k) \\ + \psi(\alpha_1, \ldots, \alpha_j - h_j, \ldots, \alpha_k)] \\ \equiv \int_{E_k'} e^{-2\pi i(\alpha, x)} \left(\sum_{j=1}^{k} (\sin \pi h_j x_j)^2 \right) dF(x) \\ \equiv \int_{E_k'} e^{-2\pi i(\alpha, x)} dF^h(x), \end{aligned}$$

where
$$F^h(A) = \int_A \left(\sum_{j=1}^{k} (\sin \pi h_j x_j)^2 \right) dF(x_j),$$

and by (3.4.11) we have

$$F^h(A) \geqq 0, \quad F^h(E_k') < \infty.$$

If now we extend $F^h(A)$ from E_k' to E_k by assigning the value 0 to the origin, then by theorem 2.1.4, $F^h(A)$ is determined uniquely in E_k and hence in E_k'. Now, for any finite number of vectors $h^\rho = (h_1^\rho, \ldots, h_k^\rho)$, $\rho = 1, \ldots, r$, this then determines uniquely the sum

$$\sum_{\rho=1}^{r} F^{h^\rho}(A) = \int_A \left(\sum_{\rho=1}^{r} (\sin \pi h_j^\rho x_j)^2 \right) dF(x),$$

but for an appropriate choice of the vectors h^ρ the integrand on the right side will be $\neq 0$ for $x \neq 0$, and thus $F(A)$ itself is uniquely determined in E_k', as claimed.

Definition 3.4.1. A family of characteristic functions

$$\phi(t;\alpha_j)=\int_{E_k} e^{-2\pi i(\alpha,x)}\,d_xF(t;x) \tag{3.4.17}$$

$$F(t;A)\geqq 0,\quad F(t;E_k)=1, \tag{3.4.18}$$

$0<t<\infty$, will be called an (*infinitely*) *subdivisible process* (or *law*) if there is a function $\psi(\alpha_j)$ such that we have

$$\phi(t;\alpha_j)=e^{-t\psi(\alpha_j)} \tag{3.4.19}$$

in all $t>0$, $\alpha\in E_k$.

Theorem 3.4.3. *A family* (3.4.17) *is of the form* (3.4.19) *if and only if we have*

$$\phi(r;\alpha)\,\phi(s;\alpha)=\phi(r+s;\alpha) \tag{3.4.20}$$

for $r>0$, $s>0$ *and*

$$\phi(t;\alpha)\overset{P}{\to}1\quad\text{as}\quad t\downarrow 0; \tag{3.4.21}$$

or, what is equivalent with it, if we have

$$F(r;A)*F(s;A)=F(r+s;A),$$

that is

$$\int_{E_n} F(r;A-y)\,d_yF(s;y)=F(r+s;A),$$

and if, furthermore, for $t\downarrow 0$, $F(t;A)$ *is Bernoulli convergent to the 'identity' distribution* J *which is concentrated all at* $x=0$, *and which can also be characterized by* $J*G=G$, *for* $G\in V$ (*or* V^+).

Proof. It is obvious that (3.4.19) does all this. Conversely, if $\phi(t;\alpha)$ is as in the theorem, then, on account of

$$\phi(r+s;\alpha)-\phi(r;\alpha)=\phi(r;\alpha)\,(\phi(s;\alpha)-1),$$

there is an open set $0<t<t_0(\alpha)$ in which we can define a continuous function $\log\phi(t;\alpha)$ which satisfies the functional equation

$$\log\phi(r+s;\alpha)=\log\phi(r;\alpha)+\log\phi(s;\alpha)$$

for $0<r,\ s<\tfrac{1}{2}t_0(\alpha)$. By known properties of this functional equation we obtain indeed $\log\phi(t;\alpha)=-t\psi(\alpha)$ for all t, α as claimed.

Theorem 3.4.4. *A function* $\psi(\alpha_j)$ *describes a subdivisible law if and only if it belongs to the closure class* (3.4.13) *that is, if and only if, it is of the form*

$$i\sum_p c_p\alpha_p+\sum_{p,q} c_{pq}\alpha_p\alpha_q+\int_{E_k'}(1-e^{-2\pi i(\alpha,x)}-2\pi i(\alpha,\lambda))\,dF(x),$$

where c_p, c_{pq}, $F(A)$ *are as described before.*

Proof. (3.4.21) implies

$$\psi(\alpha_j) = \operatorname*{P\,lim}_{u \to 0} \frac{1 - \phi(u; \alpha_j)}{u},$$

so that $\psi(\alpha_j)$ belongs to the P-closure. Conversely, if $\phi(\alpha_j)$ is a positive definite function, then

$$e^{-u[\phi(0)-\phi(\alpha)]} \equiv e^{-u\phi(0)} \sum_0^\infty \frac{u^n \phi(\alpha)^n}{n!}$$

is again positive definite, and thus $\phi(0) - \phi(\alpha)$ describes a subdivisible law. But the P-limit of subdivisible laws is again subdivisible, and hence the theorem.

3.5. Absolute moments

For any
$$\phi(\alpha) = \int_{-\infty}^{\infty} e^{-2\pi i \alpha x}\, dF(x),$$
$F \in V$, we have

$$\frac{\phi(\alpha+h) - \phi(\alpha)}{h} = \int_{-\infty}^{\infty} \frac{e^{-2\pi i h x} - 1}{hx} e^{-2\pi i \alpha x}\, x\, dF(x),$$

and since $1/hx\,(e^{-2\pi i h x} - 1)$ is in absolute value $\leqq 1$ and tends to 1 as $h \to 0$, we see that if we have

$$\int_{-\infty}^{\infty} |x|\,|dF(x)| < \infty$$

then $\phi(\alpha)$ has a continuous first derivative which can be formed under the integral. More generally, for any $m \geqq 1$ and any $k \geqq 1$, if we have

$$\int_{E_k} |x|^{2m}\, |dF(x_j)| < \infty,$$

then the Fourier transform $\phi^F(\alpha_j)$ is in $C^{(2m)}$ and the partial derivatives of order $\leqq 2m$ can be formed under the integral; so that in particular, for $k = 1$ we have

$$\frac{d^{2m}\phi(0)}{d\alpha^{2m}} = (-1)^m \int_{-\infty}^{\infty} x^{2m}\, dF(x). \qquad (3.5.1)$$

Now, it is notable that for $F(A) \geqq 0$ these assertions can be significantly inverted and it will suffice to describe the one-dimensional situation only. If we put

$$\Delta_{2m}\phi(\alpha) = \sum_{r=0}^{2m} (-1)^r \binom{2m}{r} \phi((m-r)\alpha),$$

so that in particular

$$\Delta_2 \phi(\alpha) = \phi(\alpha) - 2\phi(0) + \phi(-\alpha),$$

then we have

$$\frac{1}{\alpha^{2m}}\Delta_{2m}\phi(\alpha)=(2i)^{2m}\int_{-\infty}^{\infty}\left(\frac{\sin\pi\alpha x}{\alpha x}\right)^{2m}x^{2m}dF(x), \qquad (3.5.2)$$

and if now we replace the integral by $\int_{-A}^{A}$, for A fixed, let $\alpha\to 0$ and then $A\to\infty$, then, for $F\in V^+$, we obtain

$$\int_{-\infty}^{\infty}x^{2m}dF(x)\leq\overline{\lim_{a\downarrow 0}}\frac{\Delta_{2m}\phi(\alpha)}{\alpha^{2m}},$$

whether the quantities are finite or $+\infty$. Thus, if the function $\phi(\alpha)$ has what is called a generalized symmetric derivative of order $2m$ at the origin, or only if the corresponding difference quotient is bounded, then the $(2m)$th moment is bounded, and the function $\phi(\alpha)$ is in $C^{(2m)}$ (3.5.1) holds. Thus, for a characteristic function other than $\phi(\alpha)\equiv 1$ the $(2m)$th derivative at the origin cannot be zero for $m\geqq 1$, and this is the quickest way of verifying that for $p>2$ the function $e^{-|\alpha|^p}$ is not a characteristic function.

All this is preliminary, and it suggests that for $0<p<2m$ the relation

$$\int_{-\infty}^{\infty}|x|^p\,dF(x)<\infty$$

ought to be in some manner 'equivalent' with

$$\Delta_{2m}\phi(\alpha)=O(|\alpha|^p),$$

and we are going to show that such is indeed the case, although not literally so.

We take in $0<\alpha\leqq 1$ a measurable function $\lambda(\alpha)\geqq 0$ for which

$$\int_0^1\alpha^{2m}\lambda(\alpha)\,d\alpha<\infty,$$

and if we introduce the Poisson transform

$$\mu_0(x)=\int_0^1(\sin\pi\alpha x)^{2m}\,\lambda(\alpha)\,d\alpha,$$

multiply both sides of (3.5.2) with $\alpha^{2m}\lambda(\alpha)$ and integrate, we obtain the following conclusion:

THEOREM 3.5.1. *We have*

$$\int_0^1|\Delta_{2m}\phi(\alpha)|\,\lambda(\alpha)\,d\alpha<\infty$$

if and only if $$\int_{-\infty}^{\infty} \mu_0(x)\, dF(x) < \infty,$$

or what (due to $\mu_0(x) \leqq M_0$ in $|x| \leqq X_0$) *is the same if and only if*

$$\int_{-\infty}^{-X_0} + \int_{X_0}^{\infty} \mu_0(x)\, dF(x) < \infty.$$

From this we will conclude as follows:

THEOREM 3.5.2. *If a measurable function* $\lambda(\alpha) \geqq 0$ *in* $0 < \alpha \leqq 1$ *is such that the function*

$$\mu(x) = x^{2m} \int_0^{1/x} \alpha^{2m} \lambda(\alpha)\, d\alpha \tag{3.5.3}$$

is finite, and if $$\int_{1/x}^{1} \lambda(\alpha)\, d\alpha = O(\mu(x)), \quad \text{as} \quad x \to \infty, \tag{3.5.4}$$

then we have $$\int_{-\infty}^{-2} + \int_{2}^{\infty} \mu(x)\, dF(x) < \infty \tag{3.5.5}$$

if and only if $$\int_0^1 |\Delta_{2m} \phi(\alpha)|\, \lambda(\alpha)\, d\alpha < \infty.$$

Proof. If for $x \geqq 2$ we put $\mu_0(x) = \mu_1(x) + \mu_2(x)$, where

$$\mu_1(x) = \int_0^{1/x} (\sin \pi\alpha x)^{2m}\, \lambda(\alpha)\, d\alpha, \quad \mu_2(x) = \int_{1/x}^{1} (\sin \pi\alpha x)^{2m} \lambda(\alpha)\, d\alpha,$$

then we have $$\mu_2(x) \leqq \int_{1/x}^{1} \lambda(\alpha)\, d\alpha, \quad \mu_1(x) \leqq M_1 \mu(x)$$

and $$\mu_1(x) \geqq M_2 x^{2m} \int_0^{1/x} \alpha^{2m} \lambda(\alpha)\, d\alpha = M_2 \mu(x),$$

so that (3.5.4) implies

$$M_3 \mu(x) \leqq \mu_1(x) + \mu_2(x) \leqq M_4 \mu(x),$$

and our theorem is a consequence of the preceding one.

In particular, if we put $\lambda(\alpha) = \alpha^{-q-1} L(1/\alpha)$, where $q < 2m$ and $L(x)$ in $X_0 \leqq x < \infty$ is a so-called function of slow growth, and is monotonely decreasing, then by the theory of such functions, $\mu(x)$ is bounded above and below by $x^q L(x)$, and condition (3.5.4) is fulfilled, so that the following specific conclusion ensues:

THEOREM 3.5.3. *If* $L(x)$ *in* $X_0 \leqq x < \infty$ *is a monotonely decreasing function of slow growth, then, for* $q < 2m$, *relation*

$$\int_{-\infty}^{-1} + \int_{1}^{\infty} \mu(x)\, dF < \infty$$

is equivalent with relation

$$\int_0^1 \frac{|\Delta_{2m}\phi(\alpha)|}{\alpha^{q+1}} L\left(\frac{1}{\alpha}\right) d\alpha < \infty.$$

In particular, if for $0<p<2$ we introduce

$$f_p(x) = 2\int_0^\infty e^{-\alpha^p} \cos 2\pi x \, d\alpha$$

(probability density for a stable symmetric law), then we have

$$\int_1^\infty f_p(x)\, x^p L(x)\, dx \begin{cases} <\infty \\ =\infty \end{cases},$$

depending on whether $\displaystyle\int_0^1 \frac{1}{\alpha} L\left(\frac{1}{\alpha}\right) d\alpha \begin{cases} <\infty \\ =\infty \end{cases},$

whenever $L(x)$ is monotonely decreasing and of slow growth; and this follows from the fact that we now have

$$-\Delta_2\phi(\alpha) = 2(1-e^{-\alpha^p}) \sim 2\alpha^p$$

for small α. Thus, for instance,

$$\int_2^\infty f_p(x)\, x^p (\log x)^{-1-\epsilon}\, dx$$

is finite for $\epsilon > 0$ and infinite for $\epsilon = 0$, and similarly for the entire logarithmic scale.

3.6. Locally compact Abelian groups

If we replace the pair of dual spaces $\{E_k\colon(x), E_k\colon(\alpha)\}$ by the pair $\{T_k, M_k\}$, and if for $F \in V^+(T_k)$ we introduce the coefficients

$$\phi(m) = \int_{T_k} e^{-2\pi i(m,x)}\, dF(x), \tag{3.6.1}$$

then an analogue to theorem 3.2.4 would be vacuous because on M_k every function is continuous and the analogue to theorem 3.2.3 is as follows:

THEOREM 3.6.1. *A sequence* $\{\phi(m)\}$ *is of the form* (3.6.1) *if and only if we have*

$$\sum_{(m),(n)} \phi(m-n)\, \rho(m)\, \overline{\rho(n)} \geqq 0 \tag{3.6.2}$$

for (m) *and* (n) *ranging over any finite set in* M_k.

Proof. As previously, (3.6.2) implies

$$\sum_{(m),(n)} e^{-2\epsilon(|m+t|^2+|n+t|^2)}\phi(m-n)\,e^{-2\pi i(m-n,x)} \geqq 0,$$

and on putting $m=n+p$, the sum is

$$\sum_{(m),(p)} e^{-4\epsilon|n+t+\frac{1}{2}p|^2}e^{-\epsilon|p|^2}\phi(p)\,e^{-2\pi i(p,x)}.$$

Now, by adding over a suitable number of points t, this will be $C_\epsilon . f_\epsilon(x)$, where C_ϵ is a positive number and

$$f_\epsilon(x)=\sum_{(p)} e^{-\epsilon|p|^2}\phi(p)\,e^{-2\pi i(p,x)}.$$

Thus we have $f_\epsilon(x)\geqq 0$, and on letting $\epsilon\to 0$ we obtain the conclusion of the theorem by the use of an analogue to the closure theorem 3.2.1, easily provable.

Next, since M_k is discrete, P-convergence on it is ordinary convergence point-by-point, and the following theorem can be obtained by suitable adaptation of previous syllogisms:

THEOREM 3.6.2. *If a family $F(t; A)\in V^+(T_k)$ satisfies the assumptions of theorem 3.4.3, but on T_k, then for the Fourier coefficients*

$$\phi(t;m)=\int_{T_k} e^{-2\pi i(m,x)}\,d_x F(t;x) \tag{3.6.3}$$

we again have

$$\phi(t;m)=e^{-t\psi(m)}, \tag{3.6.4}$$

where the function $\psi(m)$ on M_k is in the P-closure of the functions

$$\psi^0(m)=\int_{T_k}(1-e^{-2\pi i(m,x)})\,dF(x), \tag{3.6.5}$$

$F\in V^+(T_k)$; and conversely any element of the P-closure gives rise to such a family $F(t; A)$.

The closure elements are again of the form

$$\psi(m)=\psi^B(m)+\psi^G(m)+\psi^P(m), \tag{3.6.6}$$

where $\psi^B(m)$ is any form $i\sum_p c_p m_p$, $\psi^G(m)$ is any P-limit of finite sums of expressions $(a_1m_1+\ldots+a_km_k)^2$, and

$$\psi^P(m)=\int_{T_k'}(1-e^{-2\pi i(m,x)}-2i\sum_p m_p\sin\pi x_p)\,dF(x), \tag{3.6.7}$$

where $T_k'=T_k-(0)$, and $F(A)\geqq 0$ is defined on T_k' and is such that

$$\int_{T_k'}\sum_p(\sin\pi x_p)^2\,dF(x_j)<\infty.$$

Also, for given $\psi(m)$ everything in the decomposition (3.6.4) *is uniquely determined.*

But now we consider the dual pair $\{M_k, T_k\}$, and in this case any element of $V^+(M_k)$ is a sequence $\{f(m)\}$ for which

$$f(m) \geqq 0, \quad \sum_{(m)} f(m) < \infty, \tag{3.6.8}$$

and its transform is

$$\phi(\alpha_j) = \sum_{(m)} e^{-2\pi i(\alpha, m)} f(m). \tag{3.6.9}$$

If we view this transform not as a function on T_k but as a periodic function on E_k, then it is a particular case of our previous transforms in E_k, for functions $F(A)$ which have nonzero values only at the lattice points, and thus theorems 3.2.3 and 3.2.4 apply readily. We also note that P-convergence on T_k is uniform convergence on T_k, due to its compactness.

However, with regard to infinitely subdivisible laws a new situation arises. If we define them suitably, then the characteristic functions

$$\phi(t; \alpha_j) = \sum_{(m)} e^{-2\pi i(\alpha, m)} f(t; m)$$

have the values $e^{-t\psi(\alpha_j)}$, and a function $\psi(\alpha_j)$ is of this kind if and only if it belongs to the P-closure of functions representable in the form

$$\sum_{(m)}{}' (1 - e^{-2\pi i(\alpha, m)}) f(m), \tag{3.6.10}$$

with (3.6.8) holding. Now, due to the fact that the origin $(m) = (0)$ is 'isolated' in the topology of M_k it can be seen, by going over previous arguments, that the class of functions (3.6.10) is already P-closed as is, so that no Gaussian distribution emerges, at least not as a limit of Poisson functions, and incidentally, no corrective term of Bernoulli term is needed either.

Actually, if we view our function (3.6.9) as a periodic function on E_k then what we have just stated implies that if an exponent (3.4.13) of a general subdivisible law in E_k happens to be multiperiodic then it is of the pristine form (3.6.10) perforce, and this assertion could have been so proven directly. But the reasoning straight on $\{M_k, T_k\}$ has the advantage that it can be generalized from M_k to any discrete Abelian group, the result being as follows.

Let G: (x) be any locally compact Abelian group and $\hat{G}$: (α) its dual group, and $\chi(\alpha; x)$ the character from G to $\hat{G}$. If with each $F(A) \in V^+(G)$ we associate the function

$$\phi(\alpha) = \int_G \chi(\alpha; x)\, dF(x), \tag{3.6.11}$$

then we take it as known from 'abstract' harmonic analysis that these functions $\{\phi(\alpha)\}$ are continuous positive-definite and P-closed. Also, if we introduce an infinitely decomposable process as before, then the exponents $\{\psi(\alpha)\}$ are again the P-closure of the set of functions

$$\psi^0(\alpha) = \int_G (1-\chi(\alpha; x))\, dF(x), \quad F \in V^+(G), \tag{3.6.12}$$

but the actual analytical representation of the closure class is not known to us, although various parts of our previous statements could be upheld by making suitable assumptions on the character function $\chi(\alpha; x)$ axiomatically. However, the following statement goes easily through, without additional assumptions being needed.

THEOREM 3.6.3. *If the Abelian group G is discrete, that is, equivalently, if $\hat{G}$ is compact then the class* (3.6.12) *is P-closed as is.*

In other words, if the values of random variables are the elements of a discrete commutative group then there are no (joint) Gaussian distributions for them possible.

3.7. Random variables

DEFINITION 3.7.1. *A (Lebesgue) measure space* is an ensemble

$$\{R; \mathscr{B}; v\}, \tag{3.7.1}$$

in which $R:(x)$ is a general point set, $\mathscr{B}:(B)$ is a σ-field of sets in R (with R itself being an element) and $v = v(B)$ is a σ-additive measure on $\mathscr{B}$, $0 \leqq v(B) \leqq +\infty$. We will call (3.7.1) *finite* if $v(R) < \infty$ and *σ-finite* if there is in $\mathscr{B}$ a sequence $B^1 \subset B^2 \subset B^3 \subset \ldots \to R$ such that $v(B^n) < \infty$, $n = 1, 2, 3, \ldots$.

If $v(R) = 1$, then, in stochastic contexts, (3.7.1) is a *probability space*, and if so viewed we will also denote it by

$$\{\Omega; \mathscr{S}; P\}, \tag{3.7.2}$$

with $\Omega = \Omega:(\omega)$, $\mathscr{S} = \mathscr{S}:(S)$, $P = P(S)$, $P(\Omega) = 1$.

DEFINITION 3.7.2. We call (3.7.1) *topological* if R is endowed with a Hausdorff topology and if $\{\mathscr{B}; v\}$ are such that for each $B \in \mathscr{B}$ and each $\eta > 0$ there is a compact set C_η in $\mathscr{B}$ such that $C_\eta \subset B$ and

$$v(C_\eta) > v(B) - \eta. \tag{3.7.3}$$

We call it *strictly topological* if every Borel set (relative to the given topology) belongs to $\mathscr{B}$.

The following fact will be taken as known.

THEOREM 3.7.1. *If R is the Euclidean E_k, for any $k \geqq 1$, and $\mathscr{B}=\mathscr{A}_k$ is the σ-field of ordinary Borel sets A, then*

$$\{E_k;\ \mathscr{A}_k;\ v\} \tag{3.7.4}$$

is strictly topological for any σ-additive $v(A)$ on $\mathscr{A}_k$.

If $v(A)$ is a probability measure then in the Euclidean case we may also denote it by $F(A)$ or $F(x_j)$ as before, and the reason for then calling it a (joint) distribution function is as follows.

Take any two spaces $R'\colon (x')$ and $R\colon (x)$ and any mapping

$$x=f(x') \tag{3.7.5}$$

from R' to all of R and for any $B \subset R$ introduce its preimage

$$B'=f^{-1}(B). \tag{3.7.6}$$

If now we are given a σ-field $\mathscr{B}$ in R then (3.7.6.) gives rise to a σ-field $\mathscr{B}^0=f^{-1}(\mathscr{B})$ in R' which we will call the *generated field*, and if a field $\mathscr{B}'$ in R' was given to start with and $\mathscr{B}^0 \subset \mathscr{B}'$ then we will call (3.7.5) a consistent mapping of $\{R', \mathscr{B}'\}$ into $\{R, \mathscr{B}\}$. Furthermore, if there is given a σ-measure $v'(B')$ on $\mathscr{B}'$, then

$$v(B)=v'(f^{-1}(B)) \tag{3.7.7}$$

defines a suchlike measure on $\mathscr{B}$, with $v(R)=v'(R')$, which we will call a *generated measure*, and if this measure $v(B)$ was so defined originally on $\mathscr{B}$ then we will call (3.7.5) a *consistent mapping* from $(R'; \mathscr{B}'; v'\}$ to $\{R;\ \mathscr{B};\ v\}$.

Take in particular for $\{R';\ \mathscr{B}';\ v'\}$ the probability space (3.7.2) and on it k real-valued Baire functions $f_1(\omega), \ldots, f_k(\omega)$, and consider the mapping

$$x_1=f_1(\omega), \ldots, x_k=f_k(\omega) \tag{3.7.8}$$

from $\Omega\colon (\omega)$ to $E_k\colon (x_j)$. Since the functions are Baire functions this is a consistent mapping from $\{\Omega;\ \mathscr{S}\}$ to $\{E_k;\ \mathscr{A}_k\}$, and the probability measure $P(S)$ then generates a probability measure $F(A) \equiv F_p(A)$ on $\mathscr{A}_k$.

DEFINITION 3.7.3. The probability measure $F(A)$ thus generated is the (joint) *distribution function* of the k functions (3.7.8), the latter functions being now called *random variables* in this context.

From Lebesgue theory we obtain the following fact:

THEOREM 3.7.2. *For any bounded Baire function* $b(x_1, \ldots, x_k)$ *in* E_k *we have*

$$\int_\Omega b(f_1(\omega), \ldots, f_k(\omega))\, dP(\omega) = \int_{E_k} b(x_1, \ldots, x_k)\, dF_p(x_j), \quad (3.7.9)$$

and, more generally, for an unbounded Baire function if one of the two integrals exists then so does the other and equality holds.

DEFINITION 3.7.4. The integral on the right-hand side in (3.7.9) is called the *expected value* of $b(x_1, \ldots, x_k)$ and is also denoted by

$$E\{b(x_1, \ldots, x_k)\} \quad \text{or} \quad E\{b(f_1, \ldots, f_k)\} \quad (3.7.10)$$

then.

It suits stochastic purposes admirably that in this last notation the awareness of the original probability space (3.7.2) is kept in abeyance and that the substitute probability space (3.7.11)

$$\{E_k;\ \mathscr{A}_k;\ F(A)\} \quad (3.7.11)$$

is being put forward instead.

For $b(x_j) = e^{-2\pi i(\alpha, x)}$, $\alpha \in E_k$, the expected value

$$\phi(\alpha_j) \equiv E\{e^{-2\pi i(\alpha, x)}\} = \int_\Omega e^{-2\pi i(\alpha_1 f_1 + \cdots + \alpha_k f_k)}\, dP(\omega) \quad (3.7.12)$$

is the characteristic function

$$\phi(\alpha_j) = \int_{E_k} e^{-2\pi i(\alpha, x)}\, dF_p(x), \quad (3.7.13)$$

but in the notation (3.7.12) it pertains to the random vector (3.7.8) rather than the set function $F(x)$.

DEFINITION 3.7.5. A sequence of random vectors

$$x_1^n = f_1^n(\omega), \ldots, x_k^n = f_k^n(\omega), \quad n = 1, 2, \ldots \quad (3.7.14)$$

is called *convergent in probability* (*or in measure*) to the vector (3.7.8) if for every bounded continuous function we have

$$E\{b(x_1^n - x_1, \ldots, x_k^n - x_k)\} \to 6(0, \ldots, 0). \quad (3.7.15)$$

For the corresponding characteristic functions we obviously obtain

$$\phi^n(\alpha_j) \overset{P}{\to} \phi(\alpha_j),$$

and for the distribution functions we have

$$F^n(A) \overset{B}{\to} F(A);$$

but these two relations do not conversely imply that the vectors (3.7.14) are convergent in probability, since distribution functions of different random vectors may be identical. What seemingly is true is that there exists some (other) probability space and on it a sequence of vectors having the same distribution functions and being convergent in probability. But this converse will not be of consequence to us (although closely related statements will) and we need not dwell on it.

Next, if we are given two probability spaces

$$\{\Omega^1: (\omega^1); \mathscr{S}^1; P^1\}, \quad \{\Omega^2: (\omega^2); \mathscr{S}^2; P^2\}, \tag{3.7.16}$$

and if (3.7.2) is their product in which

$$\omega=[\omega^1,\omega^2], \quad \Omega=\Omega^1\times\Omega^2, \quad \mathscr{S}=\mathscr{S}^1\times\mathscr{S}^2, \quad P(S^1\times S^2) = P^1(S^1).P^2(S^2),$$

then for any random variables

$$x_1=f_1(\omega^1),\ldots, \quad x_k=f_k(\omega^k), \quad y_1=g_1(\omega^2),\ldots, \quad y_l=g_l(\omega^2) \tag{3.7.17}$$

and for any two bounded Baire functions $\phi(x_1,\ldots,x_k)$, $\psi(y_1,\ldots,y_l)$ we have

$$E\{\phi(x_\kappa)\,\psi(y_\lambda)\}=E\{\phi(x_\kappa)\}.E\{\psi(y_\lambda)\} \tag{3.7.18}$$

in the sense that

$$\int_\Omega \phi(f_\kappa)\,\psi(g_\lambda)\,dP(\omega)=\int_{\Omega^1}\phi(f_\kappa)\,dP^1(\omega^1).\int_{\Omega^2}\psi(g_\lambda)\,dP^2(\omega^2),$$

and this gives rise to the following definition:

Definition 3.7.6. On *any* probability space two sets of random variables $\{x_\kappa\}$, $\{y_\lambda\}$ are (*stochastically*) *independent* if (3.7.18) holds; and more generally if a family of finite or infinite sets of random variables $\{x^\tau\}$ is given, where τ signifies membership in the family, then we call the family *independent* if for any upper indices $\tau_1,\ldots,\tau_n$, and after choosing those, for any finite subsets $x_1^{\tau_\nu},\ldots,x_{k_\nu}^{\tau_\nu}$, $\nu=1,\ldots,n$, we have

$$E\left\{\prod_{\nu=1}^{n}\phi^\nu(x_1^{\tau_\nu},\ldots,x_{k_\nu}^{\tau_\nu})\right\}=\prod_{\nu=1}^{n}E\{\phi^\nu(x_1^{\tau_\nu},\ldots,x_{k_\nu}^{\tau_\nu})\} \tag{3.7.19}$$

for any bounded Baire functions $\phi^\nu(x)$.

Given two sets of variables $\{x_\kappa\},\{y_\lambda\}$, if we introduce the characteristic function $\phi(\alpha_1,\ldots,\alpha_k)$ of the x's, the function $\psi(\beta_1,\ldots,\beta_l)$ of the y's and the function $\chi(\gamma_1,\ldots,\gamma_k,\gamma_{k+1},\ldots,\gamma_{k+l})$ of the (x,y)'s, then, in the case of independence, (3.7.18) implies

$$\phi(\alpha_1,\ldots,\alpha_k)\,\psi(\beta_1,\ldots,\beta_l)=\chi(\alpha_1,\ldots,\alpha_k,\beta_1,\ldots,\beta_l), \tag{3.7.20}$$

and it is possible to deduce from theorem 2.1.4 that the converse likewise holds.

For $l=k$, if we denote by $\rho(\delta_1,\ldots,\delta_k)$ the characteristic function of the random variables $z_1=x_1+y_1,\ldots,$ $z_k=x_k+y_k$, then theorem 2.1.1 implies

$$\phi(\alpha_1,\ldots,\alpha_k)\,\psi(\alpha_1,\ldots,\alpha_k)=\rho(\alpha_1,\ldots,\alpha_k), \tag{3.7.21}$$

and this relation is of fundamental importance indeed.

Also if we start from functions

$$\phi(\alpha_j) = \int_{E_k} e^{-2\pi i(\alpha, x)} dF^1(x_j), \psi(\beta_j) = \int_{E_k} e^{-2\pi i(\beta, y)} dF^2(y_j),$$

and if in the product

$$\phi(\alpha_j)\psi(\beta_j) = \int_{E_{2k}} e^{-2\pi i[(\alpha, x)+(\beta, y)]} dF^1(x_j)\, dF^2(y_j)$$

we put $\beta_j = \alpha_j$, then we obtain

$$\phi(\alpha_j)\psi(\alpha_j) = \int_{E_{2k}} e^{-2\pi i(\alpha, x+y)} dF^1(x_j)\, dF^2(y_j),$$

and our previous statement admits therefore of the following converse. If three characteristic functions are as in (3.7.21), then they pertain to three random vectors $x, y, x+y$ on some suitable probability space, and in fact on a $2k$-dimensional Euclidean space, as it happens.

Of this converse there are various generalizations to cases in which any (infinite) number of characteristic functions are involved, and, for instance, if we are given a set $\{e^{-t\psi(\alpha_j)}\}$ underlying a subdivisible process then on a suitable probability space there are corresponding random vectors $x = x(t)$ for which $x(r) + x(s) = x(r+s)$ is satisfied. And furthermore, if we have a decomposition $\psi(\alpha_j) = \psi^G(\alpha_j) + \psi^P(\alpha_j)$ (compare (3.4.13)), then we may secure a corresponding decomposition of the random vectors, thus, $x(t) = x^G(t) + x^P(t)$. All this is implied in a well-known theorem, but we will have a rather more general proposition of our own in which it will be contained.

3.8. General positivity

DEFINITION 3.8.1. We denote by $\mathscr{V}$ a vector space with real coefficients which is *partially ordered* in the following sense. There is defined a relation $X \geqq Y$ (or equivalently $Y \leqq X$) for some pairs of elements X, Y of $\mathscr{V}$ such that the following conditions are satisfied: (*a*) $X \geqq X$, (*b*) $X \geqq Y$, $Y \geqq X$ imply $X = Y$, (*c*) $X \geqq Y$, $Y \geqq Z$ imply $X \geqq Z$, (*d*) $X \geqq Y$ implies $X + Z \geqq Y + Z$ for any Z, (*e*) $X \geqq Y$ implies $aX \geqq aY$ for any positive number a, and (*f*) every monotone non-decreasing sequence which is bounded from above has a least upper bound. That is, $X_1 \leqq X_2 \leqq \dots \leqq X_n \leqq \dots \leqq Y$ implies the existence of an element $X' = \sup X_n$ such that $X_n \leqq X'$ $(n = 1, 2, \dots)$ and that $X_n \leqq X''$ $(n = 1, 2, \dots)$ implies $X' \leqq X''$.

We further denote by $\tilde{\mathscr{V}}$ a vector space with complex coefficients for which to every element X of $\tilde{\mathscr{V}}$ there corresponds a unique element X^* of $\tilde{\mathscr{V}}$ such that

$$(X^*)^* = X, \quad (X+Y)^* = X^* + Y^*; \quad (\rho X)^* = \bar{\rho} X^*$$

for any complex number ρ, and if $\mathscr{V}$ denotes the subset of elements for which $X^* = X$, then in $\mathscr{V}$ there is given a partial ordering as described.

We emphasize that we do not require that $\mathscr{V}$ shall be a lattice in the sense that for any two elements a 'meet' and 'join' ('sup' and 'inf') shall exist. A very significant nonlattice space $\mathscr{V}$ is the set of all bounded Hermitian operators in a Hilbert space of any dimension if we define the order relation $H_1 \leqq H_2$ to mean that $H_2 - H_1$ is positive semidefinite and if $\tilde{\mathscr{V}}$ consist of all (non-Hermitian) bounded operators then.

Theorem 3.6.1 can be generalized as follows:

THEOREM 3.8.1. *If* $\phi(m)$ *has values in* $\tilde{\mathscr{V}}$ *then the condition* (3.6.2) *is fulfilled for all finite sequences of complex numbers* $\{\rho(m)\}$ *if and only if* $\phi(m)$ *can be represented by a suitably defined Riemann integral* (3.6.1) *in which* $F(A)$ *is a finitely additive interval function whose values are non-negative elements in* $\mathscr{V}$.

It is a little harder, for general $\mathscr{V}$, $\tilde{\mathscr{V}}$, to establish the corresponding generalization of theorem 3.2.3 that a continuous function $\phi(\alpha_j)$ from E_k to $\tilde{\mathscr{V}}$ can be represented by a Fourier transform 2.1.1 for a non-negative interval function $F(A)$ with values in $\mathscr{V}$ if and only if it satisfies (3.2.17) or (3.2.18) respectively, the difficulty being simply one of finding the right version of continuity for which to secure the statement without restricting it unduly.

But for special $\mathscr{V}$, $\tilde{\mathscr{V}}$ of importance the task of setting up the theorem is sometimes quite easily accomplished. For instance if $\mathscr{V}$, $\tilde{\mathscr{V}}$ are operators in a finite dimensional Hilbert space then the elements are matrices, and thus $\phi(\alpha_j)$ is a matrix $\{\phi_{u\nu}(\alpha_j)\}$, $u, \nu = 1, \ldots, M$ and it is then not at all restrictive to assume that each component $\phi_{u\nu}(\alpha_j)$ is continuous in α by itself. This was so stipulated by Cramér and he obtained the result that such a matrix function $\phi(\alpha_j)$ is positive-definite, if and only if, component-wise, we have a representation

$$\phi_{u\nu}(\alpha_j) = \int_{E_k} e^{-2\pi i(\alpha, x)} dF_{u\nu}(x),$$

where each $F_{u\nu}(A) \in V(E_k)$ and the matrix $\{F_{u\nu}(A)\}$ is real symmetric positive semidefinite for each Borel set A.

CHAPTER 4

LAPLACE AND MELLIN TRANSFORMS

4.1. Completely monotone functions in one variable

We will frequently encounter in E_k an integral

$$\int_B \phi(t_j)\,d\rho(t_j) \qquad (4.1.1)$$

of the following description. B is a Borel set in E_k, $\phi(t_j)$ is a Baire function on B, usually continuous, and, what is important, $\phi(t_j) \geqq 0$; and $\rho(A)$ is defined on the Baire sets $A \subset B$, and is σ-additive and positive, $\rho(A) \geqq 0$. However, $\rho(A)$ may also assume the value $+\infty$, but if A is a compact subset of E_k, then $\rho(A)$ is finite. The integral (4.1.1) always 'exists' as a finite or infinite number $\geqq 0$, but whenever we introduce it as 'existing' without qualification, then we will tacitly imply that its value is a finite one.

For $k=1$, if we write down the integral

$$\int_0^\infty \phi(t)\,d\rho(t),$$

then B is the set $0 \leqq t < \infty$, and $\rho(A)$ is therefore finite for

$$A: 0 \leqq t \leqq t_0 \quad (<\infty),$$

but might become infinite if $t_0 \to \infty$. If we represent $\rho(A)$ by a monotonic point function $\rho(t)$, then $\rho(+\infty) - \rho(0) = \rho(B)$ whether this is finite or not. But if we write down the integral

$$\int_{0+}^\infty \phi(t)\,d\rho(t),$$

then B is $0 < t < \infty$, and $\rho(A)$ is finite for $a \leqq t \leqq b$, $0 < a < b < \infty$, but $\rho(A)$ may $\to \infty$ if $b \to +\infty$, or $a \to 0$, or both. For the point function $\rho(t)$ we may thus have $\rho(+\infty) = +\infty$ or $\rho(0+) = -\infty$, or both.

We take the following theorem as known:

THEOREM 4.1.1. *A function $f(x)$ in $0 < x < \infty$ has a representation*

$$f(x) = \int_0^\infty e^{-xt}\,d\rho(t), \quad \rho(A) \geqq 0 \qquad (4.1.2)$$

if and only if $f(x)$ is completely monotone in the following sense: it belongs to differentiability class $C^{(\infty)}$ and

$$(-1)^n \frac{d^n f}{dx^n} \geqq 0, \quad n=0,1,2,\dots. \tag{4.1.3}$$

The class of such functions will be denoted by CM. Actually, if we introduce the difference operator

$$\Delta_h f = f(x+h) - f(x),$$

then it suffices to demand

$$(-1)^n \Delta_{h_1}, \dots, \Delta_{h_n} f \geqq 0$$

for any $h_1 > 0, \dots, h_n > 0$ without even adding continuity, but the description used in the theorem is more pertinent to our immediate context.

DEFINITION 4.1.1. We call a continuous mapping $x = \psi(y)$ of $0 < y < \infty$ into $0 < x < \infty$ *completely monotone* if for every $f(x) \in CM$, the function $g(y) \equiv f(\psi(y))$ is also in CM.

THEOREM 4.1.2. *If $\psi(y) > 0$ and $\psi(y) \in C^{(\infty)}$ and*

$$(-1)^{n-1} \frac{d^n \psi(y)}{dy^n} \geqq 0, \quad n=1,2,3,\dots, \tag{4.1.4}$$

then the mapping $x = \psi(y)$ is completely monotone.

Proof. We have $g \geqq 0$ and $dg/dy = df/dx,\ d\psi/dy \geqq 0$ and this verifies

$$(-1)^n \frac{d^n g}{dy^n} \geqq 0, \quad n=0,1,2,\dots \tag{4.1.5}$$

for $n = 0, 1$. Now, for $n \geqq 2$ we obtain, by induction on n, an identity

$$\frac{d^n g}{dy^n} = \frac{df}{dx}\frac{d^n \psi}{dy^n} + \sum_{(p)} C^n_{p_0 p_1 \dots p_u} \frac{d^{p_0} f}{dx^{p_0}} \prod_{\mu=1}^{\nu} \frac{d^{p_\mu + 1} \psi}{dy^{p_\mu + 1}} \tag{4.1.6}$$

in which all $C^n_{(p)} \geqq 0$ and the summation extends over $p_0 \geqq 2, p_1 \geqq 0, \dots,$ $p \geqq 0$, $p_0 + p_1 + \dots + p_\nu = n$, with ν arbitrary. Relations (4.1.3) and (4.1.4) imply (4.1.5) as claimed.

There is a converse theorem that a completely monotone mapping must satisfy (4.1.4) and we will prove it in the following version.

THEOREM 4.1.3. *Let $f(x)$ be $C^{(\infty)}$ in $0 < x < \infty$, let*

$$f'(x) < 0 \quad \textit{in some interval} \quad 0 < x < x^0, \tag{4.1.7}$$

and for $n \geqq 2$ let $\quad f^{(n)}(x) = 0(f'(x)) \quad$ as $\quad x \to 0. \qquad (4.1.8)$

If $\psi(y) \in C^{(\infty)}$ and $\psi(y) > 0$ in $0 < y < \infty$, and if for every $u > 0$ the function $g_u(y) = f(u\psi(y))$ belongs to CM, that is,

$$(-1)^n \frac{d^n g_u(y)}{dy^n} \geqq 0, \quad n = 0, 1, 2, \ldots, \tag{4.1.9}$$

then $x = \psi(y)$ is a completely monotone mapping

Proof. We have by assumption (4.1.9)

$$0 \leqq -\frac{dg_u}{dy} \equiv -\frac{df}{dx} u \frac{d\psi}{dy},$$

and by assumption (4.1.7) we have, for fixed y,

$$-\frac{df}{dx} u > 0$$

for sufficiently small u and $x = u\psi(y)$. Therefore $d\psi/dy \geqq 0$, which proves (4.1.4) for $n = 1$. For $n \geqq 2$ we have

$$0 \leqq (-1)^n \frac{d^n g_u(y)}{dy^n}$$

$$\equiv (-1)^n \left[u \frac{df}{dx} \frac{d^n \psi}{dy^n} + \sum_{(p)} C^n_{p_0 p_1 \ldots p_\nu} u^{p_0} \frac{d^{p_0} f}{dx^{p_0}} \prod_{u=1}^{\nu} \frac{d^{p_\mu + 1} \psi}{dy^{p_\mu + 1}} \right],$$

and if for fixed y and small u we divide through by $-df/dx\,u$ and then let $u \to 0$, then (4.1.7) and (4.1.8) will lead to (4.1.4) as claimed.

THEOREM 4.1.4. *If $\psi(y)$ satisfies (4.1.4) and $\psi(y) > 0$, then $\psi(0+)$ exists, and we have a representation*

$$\psi(y) - \psi(0+) = \int_0^\infty \frac{1 - e^{-ty}}{t} \, d\sigma(t), \quad \sigma(A) \geqq 0, \tag{4.1.10}$$

so that also $\psi(y) = c_0 + cy + \int_{0+}^\infty (1 - e^{-ty}) \, d\chi(t), \quad \chi(A) \geqq 0, \qquad (4.1.11)$

where

$$c_0 = \psi(0+), \quad c = \sigma(0+) - \sigma(0), \quad \sigma(t) - \sigma(0+) = \int_{0+}^t t \, d\chi(t). \tag{4.1.12}$$

Conversely, if a function $\psi(y)$ in $0 < y < \infty$ can be represented in the form (4.1.11) with $c_0 \geqq 0$, $c \geqq 0$, then $y = \psi(y)$ is a completely monotone mapping.

Also, the integral in (4.1.11) is $o(y)$ as $y \to \infty$, and the representation is unique.

Proof. Relations (4.1.4) imply that $\psi'(y) \in CM$, and therefore we have

$$\psi'(y) = \int_0^\infty e^{-ty} \, d\sigma(t), \quad \sigma(A) \geqq 0,$$

and by Fubini's theorem we have

$$\psi(y)-\psi(\epsilon)=\int_0^\infty \frac{e^{-\epsilon t}-e^{-ty}}{t}\,d\sigma(t) \tag{4.1.13}$$

for $0<\epsilon<y<\infty$. Now, the integrand increases as ϵ decreases, therefore we can let $\epsilon \downarrow 0$ in (4.1.13) which leads to (4.1.10).

Conversely, if we are given (4.1.11) and if we denote the integral by $\psi_0(y)$, then we put $\psi_0(y)=\psi_1(y)+\psi_2(y)$, where

$$\left.\begin{aligned}\psi_1(y)&=\int_{0+}^{1}(1-e^{-ty})\,d\chi(t),\\ \psi_2(y)&=\int_{1+}^{\infty}(1-e^{-ty})\,d\chi(t),\end{aligned}\right\} \tag{4.1.14}$$

and examine the two parts separately. We have

$$\psi_1(1)=\int_{0+}^{1}\frac{1-e^{-t}}{t}\,t\,d\chi(t)\geqq\frac{1}{e}\int_{0+}^{1}t\,d\chi(t),$$

so that the last integral is finite. But then we can put

$$\psi_1(y)-\psi_1(0+)=\int_{0+}^{y}d\eta\int_{0+}^{1}e^{-t\eta}\,t\,d\chi(t),$$

which proves that $\psi_1'(y)$ is in CM. Also if we write

$$\frac{\psi_1(y)}{y}=\int_{0+}^{1}\frac{1-e^{-ty}}{ty}\,t\,d\chi(t)$$

and consider that $\xi^{-1}(1-e^{-\xi})$ is bounded in $0<\xi<\infty$ we can let $y\to\infty$ under the integral and we obtain $\psi_1(y)=o(y)$ as $y\to\infty$. Next, owing to

$$\frac{1}{2}\int_{1+}^{\infty}d\chi(t)\leqq\int_{1+}^{\infty}(1-e^{-t})\,d\chi(t)=\psi_2(1)<\infty,$$

it follows that we can put

$$\psi_2(y)=c_2-\int_{1+}^{\infty}e^{-ty}\,d\chi(t),$$

so that again $\psi_2'(y)\in CM$, and $\psi_2(y)$ is not only $o(y)$ but also $O(1)$ as $y\to\infty$. Finally, our reasoning implies that we have

$$\psi'(y)=c+\int_{0+}^{\infty}e^{-ty}\,t\,d\chi(t),$$

and by the uniqueness of the Laplace representation (4.1.2), which we take as known, c and $\chi(A)$ are uniquely determined and so is therefore also c_0.

For actual applications the following conclusion from the preceding theorems will be needed:

THEOREM 4.1.5. *If $\psi(x)>0$, then $e^{-u\psi(x)}$ belongs to CM for every $u>0$, if and only if $\psi'(x)$ belongs to CM.*

In particular for $0<p\leqq 1$ the function e^{-ux^p} belongs to CM for every $u>0$.

4.2. Completely monotone functions in several variables

If $k\geqq 2$ we denote by T the *closed* octant $0\leqq t_j<\infty$, $j=1,\ldots,k$, and by X the *open* octant $0<x_j<\infty$, $j=1,\ldots,k$, and we are introducing the class of functions $f(x_1,\ldots,x_k)$ in X which can be represented in the form

$$f(x_j)=\int_T e^{-(x,t)}\,d\rho(t) \qquad \rho(A)\geqq 0, \tag{4.2.1}$$

where, as always, (x,t) $x_1t_1+\ldots+x_kt_k$. Such a representation, if possible, is unique, and this can be deduced from the uniqueness of Fourier transforms as follows. If we introduce the complex variables $Z_j=x_j+iy_j$, $j=1,\ldots,k$, then the integral

$$\int_T e^{-(Z,t)}\,d\rho(t)\equiv\int_T e^{-i(y,t)}\,e^{-(x,t)}\,d\rho(t) \tag{4.2.2}$$

is absolutely uniformly convergent in the 'tube'

$$(x_1,\ldots,x_k)\in X, \quad -\infty<y_j<\infty, \quad j=1,\ldots,k,$$

and there is a holomorphic function in $Z_1,\ldots,Z_k$ whose values are uniquely determined by those of the original function $f(x_1,\ldots,x_k)$. Now, if we choose a fixed point $(x_1^0,\ldots,x_k^0)$ in X and vary the y_j's, then we can write for (4.2.2),

$$\int_{E_k} e^{-i(y,t)}\,dF(t), \tag{4.2.3}$$

where

$$F(A)=\int_{A\cap T} e^{-(x^0,t)}\,d\rho(t), \tag{4.2.4}$$

for any A in E_k. This $F(A)$ belongs to $V(E_k)$ and therefore it is uniquely determined, and since, for $A\subset T$, we have

$$\rho(A)=\int_A e^{(x^0,t)}\,dF(t),$$

it follows that $\rho(A)$ is uniquely determined too.

THEOREM 4.2.1. *A function $f(x_j)$ in an octant X can be represented by an integral* (4.2.1) *if and only if it is $C^{(\infty)}$, and we have*

$$(-1)^{n_1+\cdots+n_k}\frac{\partial^{n_1+\cdots+n_k}f}{\partial x_1^{n_1}\ldots\partial x_k^{n_k}}\geqq 0 \qquad (4.2.5)$$

for all combinations $n_1\geqq 0,\ldots,n_k\geqq 0$, the representation being unique then.

This theorem will be taken as known, and we will proceed to further developments. If for any point (ξ_j) in the point set closure $\bar{X}$ of X we introduce the operator

$$D_\xi f=\xi_1\frac{\partial f}{\partial \xi_1}+\ldots+\xi_k\frac{\partial f}{\partial \xi_k},$$

then the set of conditions (4.2.5) is equivalent to the set of conditions

$$(-1)^n D_{\xi^1}D_{\xi^2}\ldots D_{\xi^n}f\geqq 0,\quad n=0,1,2,\ldots \qquad (4.2.6)$$

for all possible points $\xi^1,\ldots,\xi^n$ in $\bar{X}$, and this new phrasing of the conditions has the following advantage. If we take any nonsingular affine transformation $x_j\to\sum_p a_{jp}x_p$, and the transposed transformation $u_j\to\Sigma\, a_{pj}u_p$—which together leave the inner product (x,t) invariant—and if we denote the corresponding images of X and T by the same letters, then the conditions (4.2.6) remain identically the same for representing a function $f(x)$ in X by an integral (4.2.1) over T. Also, the new point set X is an open 'affinely placed' octant, and T is its 'dual' octant, this one closed, and for X given T can be described as consisting of those points $t=(t_1,\ldots,t_k)$ for which

$$(x,t)\geqq 0\quad\text{for}\quad x\in X. \qquad (4.2.7)$$

If we take two affinely placed octants X_1, X_2 whose intersection $X_1\cap X_2$ is nonempty and if in $X_1\cup X_2$ we are given a function $f(x)$ which there satisfies conditions (4.2.6) for all ξ^ν in $X_1\cup X_2$, then it satisfies them separately in X_1, X_2, and thus there are two representations

$$f(x)=\int_{T_1}e^{-(x,t)}\,d\rho_1(t),\quad x\in X_1, \qquad (4.2.8)$$

$$f(x)=\int_{T_2}e^{-(x,t)}\,d\rho_2(t),\quad x\in X_2, \qquad (4.2.9)$$

where T_1 is dual to X_1 and T_2 to X_2. However, in $X_1\cap X_2$ there is some affinely placed octant X_3, $X_3 \subset X_1$, $X_3 \subset X_2$, and for its dual we have $T_3 \supset T_1$, $T_3 \supset T_2$. Thus we have a third representation

$$f(x) = \int_{T_3} e^{-(x,\,t)}\,d\rho_3(t), \quad x \in X_3, \tag{4.2.10}$$

but since $\rho_1(A)$ in T_1 can be viewed as a set function in T_3 which happens to be zero in $T_3 - T_1$, and $\rho_2(A)$ as one which is zero in $T_2 - T_1$, therefore, by the uniqueness property, a comparison of the three representations (4.2.8), (4.2.9), (4.2.10) implies that we actually have one 'joint' representation

$$f(x) = \int_{T_1 \cap T_2} e^{-(x,t)}\,d\rho(t)$$

for x in $X_1 \cup X_2$.

From this we can conclude that if we are given a family of affinely-placed octants $\{X_\alpha\}$ in which any two can be connected by a finite chain in which two successive ones overlap, and if in $X = \bigcup_\alpha X_\alpha$ we are given a function $f(x_j)$ satisfying (4.2.6), then it can be represented by the integral (4.2.1) in which T is the intersection $\bigcap_\alpha T_\alpha$. Also, T can be defined purely in terms of the set X by (4.2.7), and it always contains at least the origin $(0, \ldots, 0)$. Now, if the integral (4.2.1) converges in a point set X it also converges in its convex hull, and the point set T as defined by (4.2.7) remains the same. For our point set X, the hull is an open set, and, if it contains the origin it contains the entire E_k and the point set T must consist of the origin only, so that $f(x_j)$ is a constant perforce. If, however, the hull does not contain the origin, then it is itself a connected union of octants, and it is a 'cone' in the sense of the following definition:

DEFINITION 4.2.1. *A cone* is an open convex set not containing the origin which is a union of affinely-placed octants or equivalently a union of half-lines $\{\rho x_1, \ldots, \rho x_k\}$, $(x_1, \ldots, x_k) \in X$, $0 < \rho < \infty$. We call it a *proper* cone if it is part of an affinely-placed octant, that is, if after an affine transformation centred at the origin it becomes part of the octant $x_1 > 0, \ldots, x_k > 0$.

A cone is a proper cone if and only if its dual set T as defined by (4.2.7) contains a neighborhood, in which case T is itself the closure of an (open) proper cone. Otherwise, T is contained in a linear subspace through the origin, and has this structure there, or it reduces to the origin.

Returning to the function $f(x)$ we may now summarize as follows:

THEOREM 4.2.2. *A function $f(x_j)$ in a cone X has a representation* (4.2.1) *there if and only if it is $C^{(\infty)}$ and satisfies relations* (4.2.6) *there.*

These functions will be called *completely monotone* and their class will be denoted by $CM(X)$.

DEFINITION 4.2.2. If X is a cone in E_k: (x_p) and Y is a cone in E_l: (y_q), $k \geqq 1$, $l \geqq 1$, $k \gtreqless l$, and if

$$x_p = \psi_p(y_1, \ldots, y_l), \quad p = 1, \ldots, k \tag{4.2.11}$$

is a transformation from Y to X, then we call it a *completely monotone mapping* if $\psi_p \in C^{(\infty)}$, and if for any points $\eta^1, \ldots, \eta^n$ in Y, and any y in Y, the point in E_k with the coordinates

$$x_p = (-1)^n D_{\eta^1} \ldots D_{\eta^n} \psi_p, \quad p = 1, \ldots, k$$

lies in the closure $\bar{X}$ of X.

THEOREM 4.2.3. *If $f(x) \in CM(X)$, and* (4.2.11) *is a completely monotone mapping, then $g(y) \equiv f(\psi_1, \ldots, \psi_k) \in CM(Y)$.*

Conversely, if (4.2.11) *is a transformation $C^{(\infty)}$ from Y to X, if X is a proper cone and if for every $t \in T$, which is the dual to X, the function*

$$e^{-(t_1\psi_1(y) + \cdots + t_k\psi_k(y))} \tag{4.2.12}$$

is in $CM(Y)$, then (4.2.11) *is a completely monotone mapping.*

The proofs, though outwardly more elaborate, are the same as for $k = 1$ and we will omit them.

THEOREM 4.2.4. *A single function $x = \psi(y_1, \ldots, y_l)$ in Y is a completely monotone mapping from Y to the one-dimensional cone $0 < x < \infty$ if and only if we can put*

$$\psi(y_j) = c_0 + \sum_q c_q y_q + \int_{U'} (1 - e^{-(y,u)})\, d\sigma(u) \tag{4.2.13}$$

where $c_0 \geqq 0$, $c_q \geqq 0$, $\sigma(A) \geqq 0$, and U' is the point set arising from the dual U of Y by deletion of the origin; the representation (4.2.13) *being unique then.*

If X is a proper cone, then a transformation (4.2.11) *from Y to X is a completely monotone mapping, if and only if we can put*

$$\psi_p(y_j) = c_{p0} + \sum_q c_{pq} y_q + \int_{U'} (1 - e^{-(y,u)})\, d\sigma_p(u) \tag{4.2.14}$$

where c_{p0}, c_{pq} are real numbers and each $\sigma_p(A)$ is the difference of two *non-negative set functions on U', $\sigma_p(A)=\sigma_p^+(A)-\sigma_p^-(A)$, such that for each $(t_1, \ldots, t_k)$ in T the combination*

$$\sum_p t_p\psi_p(y) \equiv \sum_p t_p c_{p0} + \sum_{pq} t_p c_{pq} y_q + \int_{U'} (1-e^{-(y,u)})\, d\left(\sum_p t_p d\sigma_p(u)\right)$$

is as in the first part of the theorem.

The second part of the theorem can be easily obtained from definition 4.2.2. As for the first part of the theorem we note that the cone Y contains an affinely-placed octant Y_0, so that correspondingly we have $U \subset U_0$, and it is not hard to see that we need only continue with the proof for the case Y: $\{0<y_q<\infty\}$, U: $\{0 \leqq u_q < \infty\}$.

Now, a function $x=\psi(y_q)$ is a completely monotone *mapping* of this Y into $0<x<\infty$ if and only if the l functions

$$\frac{\partial\psi}{\partial y_q}, \quad q=1, \ldots, l \tag{4.2.15}$$

exist and are a completely monotone *function* each, and we need only analyze functions of this kind.

Now, if we are given (4.2.13), then it is not hard to generalize the reasoning used for $l=1$ and show that $\psi(y_j)$ has first partial derivatives, which can be obtained thus:

$$\frac{d\psi}{dy_q} = c_q + \int_{U'} e^{-(y,u)} u_q d\sigma(u),$$

and this not only implies that the functions (4.2.15) are in $CM(Y)$ but that everything is uniquely determined too.

The converse is less obvious. Since the functions (4.2.15) are in $CM(Y)$ we have representations

$$\frac{\partial\psi}{\partial y_q} = c_q + \int_{U'} e^{-(y,u)} d\sigma_q(u), \quad q=1, \ldots, l$$

with certain set functions $\sigma_q(A) \geqq 0$. Now, if we differentiate this with respect to y_r and interchange (q, r) and compare, we obtain

$$u_r d\sigma_q(u) = u_q d\sigma_r(u), \tag{4.2.16}$$

$q, r = 1, \ldots, l$, in the sense that

$$\int_A u_r d\sigma_q(u) = \int_A u_q d\sigma_r(u) \tag{4.2.17}$$

for every $A \subset U'$. For fixed r, if we denote by A_r the subset of U' for which $u_r=0$, then (4.2.17) implies

$$\int_{A_r} u_q d\sigma_r(u) = 0, \quad q=1, 2, \ldots, l.$$

Hence $$\int_{A_r} (u_1+\ldots+u_l)\, d\sigma_r(u) = 0,$$
but in U' we have $u_1+\ldots+u_l>0$. Therefore we have

$$\int_{A_r} d\sigma_r(u) = 0 \quad \text{for} \quad r=1,\ldots,l,$$

and from this we can conclude that in (4.2.16) we can divide out thus:

$$\frac{d\sigma_q(u)}{u_q} = \frac{d\sigma_r(u)}{u_r}, \quad q,r=1,\ldots,l,$$

in the sense that there exists a set function $\sigma(A)\geqq 0$, such that

$$\int_A d\sigma_q(u) = \int_A u_q\, d\sigma(u), \quad q=1,\ldots,l$$

for $A \subset U'$. Therefore, we have

$$\frac{\partial\psi}{\partial y_q} = c_q + \int_{U'} e^{-(y,\,u)}\, u_q\, d\sigma(u), \tag{4.2.18}$$

$c_q \geqq 0$, $\sigma(A)\geqq 0$, and if for some fixed $\eta_1>0,\ldots,\eta_l>0$ we introduce the function
$$\psi_\eta(y) = \psi(\eta_1 y, \ldots, \eta_l y),$$

then $$\frac{\partial \psi_\eta}{\partial y} = \sum_q c_q \eta_q + \int_{U'} e^{-y(\eta,\,u)}\, (\eta, u)\, d\sigma(u),$$

and by a judicious application of theorem 4.1.4 we now obtain

$$\psi_\eta(y) = \psi_\eta(0+) + \Sigma c_q \eta_q y + \int_{U'} (1-e^{-y(\eta,\,u)})\, d\sigma(u).$$

If now we put $y=1$ and vary the $\eta_1,\ldots,\eta_l$, denoting them by $y_1,\ldots,y_l$, we conclude that the function

$$\Sigma c_q y_q + \int_{U'} (1-e^{-(y,\,u)})\, d\sigma(u)$$

is finite, and since its derivatives have the values (4.2.18) it differs from ψ by a constant c_0, and since this constant has the value $\psi_y(0+)$ it is $\geqq 0$, as claimed.

4.3. Subordination of infinitely subdivisible processes

If a completely monotone function in $0<y<\infty$

$$\mu(y) = \int_0^\infty e^{-ty}\, d\gamma(t) \tag{4.3.1}$$

is bounded there, say $\mu(0+) \equiv \gamma(+\infty)-\gamma(0)=1$, and if we introduce the complex variable $\eta = y+iy'$, then the integral converges in the closed half-plane $0 \leqq y < \infty$, $-\infty < y' < \infty$, and uniformly in every compact set of it, and the resulting function which is holomorphic in the open half-plane will be denoted by $\mu(\eta)$. Likewise, a completely monotone mapping from $y > 0$ to $z > 0$,

$$\nu(y) = cy + \int_{0+}^{\infty} (1 - e^{ty})\, d\rho(t) \equiv cy + \nu_0(y) \tag{4.3.2}$$

can be extended in the same manner into the closed half-plane, as can be seen from the decomposition (4.1.14). Now, for any of our 'exponents' $\psi(\alpha_j)$ in E_k we have $\operatorname{Re} \psi(\alpha_j) \geqq 0$, and thus we can form

$$\tilde{\psi}(\alpha_j) = \nu(\psi(\alpha_j)), \tag{4.3.3}$$

for which again $\operatorname{Re} \tilde{\psi}(\alpha_j) \geqslant 0$. Since $\operatorname{Re} \nu_0(\eta) \geqq 0$ and $\nu_0(\eta)$ is analytic in $y > 0$, it follows that we cannot have $\operatorname{Re} \nu_0(\eta) = 0$ at an interior point with $y > 0$ unless $\nu_0((\eta) \equiv 0$; and if we have $\operatorname{Re} \nu_0(\eta) = 0$ for a boundary point $\eta = iy'$,

$$\int_{0+}^{\infty} (1 - \cos ty')\, d\rho(t) = 0,$$

then this implies that the σ-additive function $\rho(A)$ must be 0 on every Borel set A not containing a point $t = 2\pi n/y'$, and thus we must have $\operatorname{Im} \nu_0(iy') = 0$ likewise. In particular, $\operatorname{Re} \nu_0(\psi(\alpha_j)) = 0$ implies

$$\nu_0(\psi(\alpha_j)) = 0$$

for any $\alpha \in E_k$.

THEOREM 4.3.1. *If $\{e^{-t\psi(\alpha_j)}\}$ is any subdivisible process, and if we form* (4.3.3) *with any* (4.3.2), *then $\{e^{-t\tilde{\psi}(\alpha)}\}$ is again such a process, and we call $\{\tilde{\psi}(\alpha_j)\}$ 'subordinate' to $\{\psi(\alpha_j)\}$.*

Also, a subordinate process of a subordinate process is again subordinate.

Proof. Since $\mu(y; u) \equiv e^{-u\nu(y)}$ is an element of CM for every $u \geqq 0$, we have

$$\mu(y; u) = \int_{0}^{\infty} e^{-ry}\, d_r \gamma(r; u) \tag{4.3.4}$$

with $d_r \gamma(r; u) \geqq 0$, and hence

$$e^{-u\tilde{\psi}(\alpha)} = \int_{0}^{\infty} e^{-r\psi(\alpha)}\, d_r \gamma(r; u). \tag{4.3.5}$$

Now, the left side in (4.3.5) is obviously continuous in α, and the integral obviously satisfies the condition

$$\int_{0}^{\infty} \left(\sum_{p,q=1}^{N} e^{-r\psi(\alpha^p - \alpha^q)} \rho_p \rho_q \right) d_r \gamma(r; u) \geqq 0$$

if the integrand does, and thus Theorem 3.2.3 can be applied. Also, the 'chain rule' for subordination is a consequence of the corresponding chain rule for completely monotone mappings, which in its turn follows immediately from definition 4.1.1.

We now put

$$\psi(\alpha) = \psi^G + \psi^R + i\psi^I, \quad \tilde{\psi} = \tilde{\psi}^G + \tilde{\psi}^R + i\tilde{\psi}^I,$$

where ψ^G is the Gaussian addend as in (3.4.13) and ψ^R, ψ^I are the real and imaginary parts of $\psi^B + \psi^P$. Relation (4.3.2) leads then to

$$\tilde{\psi}^G + \tilde{\psi}^R = c(\psi^G + \psi^R) + \operatorname{Re} \nu_0(\psi^G + \psi^R + i\psi^I), \tag{4.3.6}$$

$$\tilde{\psi}^I = c\psi^I + \operatorname{Im} \nu_0(\psi^G + \psi^R + i\psi^I), \tag{4.3.7}$$

but since $\tilde{\psi}^R$, ψ^R and $\nu_0(\tilde{\psi})$ are all $o(|\alpha|^2)$ as $|\alpha| \to \infty$, by Theorems 3.4.2 and 4.1.4, relation (4.3.6) can be separated into

$$\tilde{\psi}^G = c\psi^G, \tag{4.3.8}$$

$$\tilde{\psi}^R = c\psi^R + \operatorname{Re} \nu_0(\psi^G + \psi^R + i\psi^I). \tag{4.3.9}$$

All three terms in 4.3.9 are $\geqq 0$, and $\operatorname{Re} \nu_0 = 0$ implies also $\operatorname{Im} \nu_0 = 0$, and hence if we assume $\tilde{\psi}^R = 0$, and then write $\psi(\alpha)$ instead of $\tilde{\psi}(\alpha)$, we are led to the following major conclusion:

THEOREM 4.3.2. *A subdivisible process with an exponent of the form*

$$\psi(\alpha) = \psi^G(\alpha) + i\psi^I(\alpha) \tag{4.3.10}$$

is not subordinate to any subdivisible process whatsoever but itself.

The 'stable laws' $\quad \{e^{-t|\alpha|^{2p}}\}, \quad 0 < p < 1 \qquad$ (4.3.11)

are each subordinate to the Gaussian law $\{e^{-t|\alpha|^2}\}$, and quite generally if $\{e^{-t\psi(\alpha)}\}$ is any subdivisible process then so is $\{e^{-t\psi(\alpha)^p}\}$ for any $0 < p < 1$. Also, the relations (4.3.5) imply for the joint distribution functions the relations

$$\tilde{F}(u; A) = \int_{r=0}^{\infty} F(r; A)\, d_r\gamma(r; u), \tag{4.3.12}$$

whose precise manner and range of validity will still be discussed, and they state that $\tilde{F}(u; A)$ is a certain average over the set function $F(r; A)$, the averaging weight being '$d_r\gamma(r; u)$'. Also, the averaging process may be viewed either as a mixing of coexisting probabilities, or as a randomization of the parameter r in alternate probabilities. We might state though that the stable and similar subordinate laws are usually arrived at in the central limit theory, and there the objects of study are families of random variables on a common ensemble $\{\Omega; \mathscr{S}; P\}$, and not families of probability measures on a common ensemble $\{\Omega; \mathscr{S}\}$.

THEOREM 4.3.3. *A process* $\{\psi(\alpha_j)\}$ *is subordinate to a Gaussian process* $\left\{Q(\alpha_j) \equiv \sum_{p,q=1}^{k} c_{pq}\alpha_p\alpha_q\right\}$ *if and only if we have*

$$\psi^G(\alpha_j) = cQ(\alpha_j) \tag{4.3.13}$$

and

$$\psi^P(\alpha_j) = \int_{E'_k} (1 - \cos(\alpha, x)) f(Q'(x_j))\, dv_x, \tag{4.3.14}$$

where $f(y)$ *is a completely monotone function in* $0 < y < \infty$ *and* $Q'(x_j)$ *is the quadratic form inverse to* $Q(\alpha_j)$.

Proof. By subordination we have

$$\psi(\alpha_j) = cQ(\alpha_j) + \int_{0+}^{\infty} (1 - e^{-tQ(\alpha_j)})\, d\rho(t), \tag{4.3.15}$$

and by the theorem 3.4.4 we have

$$\psi(\alpha_j) = \psi^G(\alpha_j) + \int_{E'_k} (1 - \cos(\alpha, x))\, dF(x), \tag{4.3.16}$$

and since by theorems 4.1.4 and 3.4.2 the two integrals are both $o(|\alpha|^2)$ we hence obtain relation (4.3.13) and also

$$\psi^P(\alpha) \equiv \int_{E'_k} (1 - \cos(\alpha, x))\, dF(x) = \int_{0+}^{\infty} (1 - e^{-tQ(\alpha_j)})\, d\rho(t). \tag{4.3.17}$$

We will now make use of the formula

$$1 - e^{-tQ(\alpha_j)} = \frac{c_0}{t^{\frac{1}{2}k}} \int_{E_k} (1 - \cos(\alpha, x))\, e^{-(1/t)Q'(x_j)}\, dv_x \tag{4.3.18}$$

where $Q'(x_j)$ is the inverse to $Q(\alpha_j)$ except for a factor $c_1 > 0$, but since $f(c_1 y)$ is completely monotone if $f(y)$ is it will be justified to act as if we had $c_1 = 1$. Now, if we substitute (4.3.18) in the second integral (4.3.17) we obtain (4.3.14) with

$$f(y) = \int_{0+}^{\infty} e^{-y/t} \frac{c_0}{t^{\frac{1}{2}k}}\, d\rho(t),$$

and for this we can write $\displaystyle\int_{0+}^{\infty} e^{-\tau y}\, d\sigma(\tau)$,

where $d\sigma(\tau) = c_0 \tau^{\frac{1}{2}k} d\left[-\rho\left(\frac{1}{\tau}\right)\right]$. Conversely, if we start from (4.3.14) for any completely monotone $f(y)$, we can gain back the second integral (4.3.17), q.e.d.

Turning to spaces other than E_k, we first of all note that the statements in section 3.6 can be supplemented as follows:

THEOREM 4.3.4. *Our definition of subordination and theorem* 4.3.1 *also apply to locally compact Abelian groups.*

Thus, in particular, if we are given any process

$$e^{-r\psi(m)} = \int_{T_k} e^{-2\pi i(m,x)} d_x F(r;x), \tag{4.3.19}$$

and if with any function (4.3.2) *we form the exponents*

$$\tilde{\psi}(m) = \nu(\psi(m)) \tag{4.3.20}$$

then there is a process

$$e^{-u\tilde{\psi}(m)} = \int_{T_k} e^{-2\pi i(m,x)} d_x \tilde{F}(u;x), \tag{4.3.21}$$

and in this particular case, theorems 4.3.2 *and* 4.3.3 *likewise apply.*

Now, our 'subordination' can also be introduced on spaces in general provided we shift the emphasis from Fourier transformation (which might not even be definable) to (generalizations of) the distributions $F(u;A)$, and we are going to do this now rather systematically.

4.4. Subordination of Markoff processes

Definition 4.4.1. Any function $\gamma(r;u)$ in $0 \leqq r < \infty$, $0 \leqq u < \infty$ which occurs in a formula (4.3.4) will be called a *subordinator*, and it will be called a *proper subordinator* if

$$\gamma(0+;u) - \gamma(0;u) = 0 \quad \text{for} \quad u > 0. \tag{4.4.1}$$

Theorem 4.4.1. *A function $\gamma(r;u)$ in $r \geqq 0$, $u \geqq 0$ is a subordinator if and only if* (i) *it is monotone in r and $\gamma(\infty;u) - \gamma(0;u) = 1$, and $\gamma(0+;0) - \gamma(0;0) = 1$,* (ii) *we have*

$$\int_{r+s=t} d_r\gamma(r;u)\, d_s\gamma(s;v) = d_t\gamma(t;u+v) \tag{4.4.2}$$

in the sense that

$$\gamma(t;u+v) - \gamma(0;u+v) = \int_{0 \leqq s \leqq t} (\gamma(t-s;u) - \gamma(0;u))\, d_s\gamma(s;v)$$

('*additivity*'), *and* (iii) *for each $r_0 > 0$ we have*

$$\lim_{u \downarrow 0} \int_{r_0}^{\infty} d\gamma(r;u) = 0, \tag{4.4.3}$$

that is,

$$\lim_{u \downarrow 0} \int_0^{r_0} d\gamma(r;u) = 1. \tag{4.4.4}$$

Also, if (4.4.1) *holds for one $u_0 > 0$ it holds for all, and it so holds if and only if $\nu(y) \to \infty$ as $y \to \infty$.*

Proof. For a subordinator, (i) is obvious, (ii) follows from

$$e^{-u\nu(y)}\, e^{-v\nu(y)} = e^{-(u+v)\nu(y)}$$

and (iii) from

$$1-e^{-u\nu(1/r_0)}=\int_0^\infty (1-e^{t/r_0})\,d\gamma(t;\,u)\geqq\int_{r_0}^\infty (1-e^{-t/r_0})\,d\gamma(t;\,u)$$
$$\geqq\left(1-\frac{1}{e}\right)\int_{r_0}^\infty d\gamma(t;\,u).$$

Conversely, if (i) holds we can set up (4.3.4), and if we write it as $e^{-\nu(y;\,u)}$ with $\nu(y;\,u)\geqq 0$, then (ii) implies $\nu(y;\,u+v)=\nu(y;\,u)+\nu(y;\,v)$. Also

$$\mu(y;\,u)\geqq\int_0^{r_0} e^{-yt}\,d\gamma(t;\,u)\geqq e^{-yr_0}\int_0^{r_0} d\gamma(t;\,u),$$

and (iii) implies $\lim\nu(y;\,u)=0$ as $u\to 0$, and therefore $\nu(y;\,u)=u\nu(y)$, as claimed.

Finally, (4.4.1) is equivalent with $\lim\limits_{y\to\infty}\mu(y;\,u)=0$, that is, $u\nu(y)\to\infty$ as $y\to\infty$, and this completes the proof of the theorem.

THEOREM 4.4.2. *In E_k (and perhaps in all locally compact commutative groups) the precise meaning of relation* (4.3.12) *is that we have*

$$\int_{E_k} c(x)\,d_x\tilde{F}(u;\,x)=\int_0^\infty\left(\int_{E_k} c(x)\,d_xF(r;\,x)\right)d_r\gamma(r;\,u) \tag{4.4.5}$$

for all continuous $c(x)$ which are 0 at infinity or any subclass as in lemma 1.5.1.

Proof. If $\chi(\alpha)\in L_1(E_k)$, then, as is easily seen, (4.3.5) implies

$$\int_{E_k}\chi(\alpha)\,e^{-u\tilde{\psi}(\alpha)}\,dv_\alpha=\int_0^\infty\left(\int_{E_k}\chi(\alpha)\,e^{-r\psi(\alpha)}\,dv_\alpha\right)d_r\gamma(r;\,u), \tag{4.4.6}$$

and conversely if relation (4.4.6) holds for $\chi(\alpha)$ variable and other entities fixed then (4.3.5) holds. Therefore, (4.3.5) is equivalent with relation (4.4.5) for functions $c(x)$ which are Fourier transforms of functions in $L_1(E_k)$. But any continuous $c(x)$ which is 0 at infinity is a uniform limit of such ones, and it is easily seen that (4.4.5) remains valid then.

For $r=0$, $F(r;\,A)$ is the 'identity', and not absolutely continuous, but for $r>0$ we may have

$$F(r;\,A)=\int_A f(r;\,x)\,dv_x, \tag{4.4.7}$$

$f(r;\,x)\geqq 0$; and if this is so then the right side in (4.4.5) can be written as

$$\int_0^\infty\left(\int_{E_k} f(r;\,x)\,c(x)\,dv_x\right)d_r\gamma(r;\,u), \tag{4.4.8}$$

provided $\gamma(r; u)$ is a *proper* subordinator. Also, if $f(r; x)$ is a Baire function in (r, x) say, then $\tilde{F}(u; x)$ is likewise an integral of a suitable function $\tilde{f}(u; x)$ such that for every u we have

$$\tilde{f}(u; x) = \int_0^\infty f(r; x) \, d_r \gamma(r; u) \tag{4.4.9}$$

for almost all x; but it does not follow that $\tilde{f}(u, x)$ can be also assumed to be a Baire function in (u, x) automatically. Furthermore, apart from subordination, the continuity of $F(r; x)$ at $r = 0$ can be expressed by the condition

$$\lim_{r \downarrow 0} \int_{Ek} f(r; x - y) \, c(y) \, dv_y = c(x) \tag{4.4.10}$$

as well.

We are now going to envisage spaces in general but in order to avoid a number of complications we will generalize the densities $f(r; x)$ only, and not also the set functions $F(u; A)$ in general.

DEFINITION 4.4.2. Given a measure space

$$\{R; \mathscr{B}; v\}, \tag{4.4.11}$$

a *Markoff* (*chain*) *density* is a function $f(r, s; x, y)$ which is defined for $0 \leqq r < s < \infty$; $x, y \in R$ and has the following properties. It is a Baire function in (r, s, x, y) and

$$f(r, s; x, y) \geqq 0, \qquad \int_R f(r, s; x, y) \, dv_y = 1 \tag{4.4.12}$$

and

$$\int_R f(r, s; x, y) f(s, t; y, z) \, dv_y = f(r, t; x, z), \tag{4.4.13}$$

the integral existing without exceptions.

We call the density *continuous* if we have

$$\lim_{s \downarrow r} \int_R f(r; s; x, y) \, c(y) \, dv_y = c(x) \tag{4.4.14}$$

for $c(y) \in C$, where C is a given set of bounded Baire functions having the completeness property that for $g(y) \in L_1(R)$ we can have

$$\int_R g(y) \, c(y) \, dv_y = 0$$

for all $c(y) \in C$ only if $g(y) = 0$ a.e.

We call it *time homogeneous* or more frequently 'stationary' if there is a Baire function $f(t; x, y)$ for $0 < t < \infty$; $x, y \in R$, such that

$$f(r, s; x, y) = f(s - r; x, y),$$

the new function having the properties

$$f(t;\,x,y)\geqq 0,\quad \int_R f(t;\,x,y)\,dv_y=1 \tag{4.4.15}$$

$$\int_R f(r;\,x,y)f(s;\,y,z)\,dv_y=f(r+s;\,x,z), \tag{4.4.16}$$

with (4.4.14) being now

$$\lim_{t\downarrow 0}{}' \int_R f(t;\,x,y)\,c(y)\,dv_y=c(x). \tag{4.4.17}$$

We call it *space homogeneous* if there is given on R a fixed transitive group of point transformations $x'=Ux$ of R onto itself such that

$$U\mathscr{B}=\mathscr{B},\quad v(UB)=v(B),\quad f(r,s;\,Ux,Uy)=f(r,s;\,x,y). \tag{4.4.18}$$

We call it (*fully*) *homogeneous* if it is homogeneous both in time and in space.

The reader will easily verify the following statement:

THEOREM 4.4.3. *If $f(t;\,x,y)$ is a stationary density, and if for a proper subordinator the integral*

$$\tilde{f}(u;\,x,y)=\int_0^\infty f(r;\,x,y)\,d_r\gamma(r;\,u) \tag{4.4.19}$$

is Baire in (u,x,y) then it is again such a density, and we call it subordinate to the original one. (A subordinate of a subordinate is again subordinate.)

Also, if $f(t;\,x,y)$ is space homogeneous or continuous then so is also $\tilde{f}(u;\,x,y)$ respectively.

Note that if (4.4.11) is the 'ordinary' Lebesgue space $\{E_k;\,\mathscr{A}_k;\,v\}$, then (4.4.18) holds for the group of translations $x_j'=x_j+a_j,\,j=1,\ldots,k$, and in this case space homogeneity of a density means that there is a function $f(r,s;\,z_j)$ such that $f(r,s;\,x,y)=f(r,s;\,y_j-x_j)$. Also, if stationarity is added then there exists a function $f(t;\,z_j)$ such that $f(t;\,x,y)=f(t;\,y_j-x_j)$, and in the continuous case this is then the function occurring in (4.4.7) under the integral. Similarly for locally compact Abelian groups in general.

Now, the group of translations is not the largest group for which (4.4.18) holds, not even among continuous transformations, the largest among the latter ones being the group of motions which includes orthogonal transformations as well. Now, for the latter group the functions $f(r,s;\,z_j)$ and $f(t;\,z_j)$ depend on the distance $|z|$ only, and the Fourier transforms of such continuous $f(t;\,z_j)$ are the functions

$e^{-t\psi(\alpha_j)}$ in which $\psi(\alpha_j)$ depends likewise on $|\alpha|$ only. Now, such a $\psi(\alpha_j)$ is real-valued since the imaginary part changes its algebraic sign under the orthogonal transformation $a'_j = -\alpha_j$, and, by uniqueness, the Gaussian and Poisson parts of $\psi(\alpha_j)$ must be radial separately. Here we have

$$\psi(\alpha_j) = c\,|\,\alpha\,|^2 + \int_{E'_k} (1 - \cos(\alpha, x))\, dF(x), \tag{4.4.20}$$

where $F(UA) = F(A)$ for any rotation around the origin. Now, if we 'radialize' the integral, and denote by $G(R)$ a function for which $G(R) - G(r)$ is the value of $F(A)$ for $r \leqslant |\,x\,| < R$, then we can write for it

$$\int_{0+}^{\infty} (1 - H_{\frac{1}{2}(k-2)}(|\,\alpha\,|\,.R)\, dG(R)),$$

where $H_\nu(z)$ is the function $c_\nu J_\nu(z)\, z^{-\nu}$, with c_ν being so chosen that $H_\nu(0) = 1$. Therefore

$$H_\nu(z) = \sum_{n=0}^{\infty} (-1)^n \frac{\Gamma(\nu+1)}{\Gamma(n+\nu+1)} (\tfrac{1}{2}z)^{2n} \frac{1}{n!},$$

and if we let $k \to \infty$, that is, $\nu \to \infty$, then $H_\nu(z) \to e^{-\frac{1}{4}z^2}$. Therefore, if we put $t = R^2$, $G(R) = \gamma(t)$, the formal limit of (4.4.20) is

$$\nu(|\,\alpha\,|^2) = c\,|\,\alpha\,|^2 + \int_{+0}^{\infty} (1 - e^{-\frac{1}{4}(|\,\alpha\,|^2)t})\, d\gamma(t) \tag{4.4.21}$$

and actually the following theorem can be proven:

THEOREM 4.4.4. *If a function $\nu(y)$ is such that $\nu(|\,\alpha\,|^2)$ describes a subdivisible process for $|\,\alpha\,|^2 = \alpha_1^2 + \ldots + a_k^2$, for all $k \geqq 1$, then $\nu(y)$ must be of the form* (4.3.2).

4.5. A theorem of Hardy, Littlewood and Paley

DEFINITION 4.5.1. Given (4.4.11) we call a function $f(x, y)$ *symmetric*, if $f(x, y) = f(y, x)$, and we call $\{f(t; x, y)\}$ symmetric if it is so for each t.

Note that in E_k or T_k if $f(x, y) = f(x - y)$ and $f(z) \in L_1$, then $f(x, y)$ is symmetric if and only if $\phi_f(\alpha)$ is real-valued.

THEOREM 4.5.1. *If $f(x, y)$ is symmetric and*

$$f(x, y) \geqq 0, \quad \int f(x, y)\, dv_y = 1 = \int f(x, y)\, dv_x, \tag{4.5.1}$$

and if we introduce the distributive transformation

$$h(x) = \int_R f(x, y)\, g(y)\, dv_y, \tag{4.5.2}$$

then for $g(y) \geqq 0$ we have

$$\int h(x)^p \, dv_x \leqq \int g(y)^p \, dv_y \tag{4.5.3}$$

for every $p \geqq 1$, and for $p = 1$ equality holds.

Proof. Obviously

$$\int h(x) \, dv_x = \int \left(\int f(x,y) \, dv_x \right) g(y) \, dv_y = \int g(y) \, dv_y$$

as claimed for $p = 1$, and for $p \geqq 1$, if we apply Holder's inequality to $f(x,y)^{1/q} . f(x,y)^{1/p} g(y)$ we obtain first

$$h(x)^p \leqq \left(\int f(x,y) \, dv_y \right)^{p/q} \int f(x,y) \, g(y)^p \, dv_y = \int f(x,y) \, g(y)^p \, dv_y,$$

and if we now integrate with respect to x we obtain (4.5.3).

Note that only assumptions (4.5.1) have been used and not the symmetry in its entirety.

DEFINITION 4.5.2. We say that a Markoff density $\{f(t; x, y)\}$ is of *special kind* if it is symmetric and if there is a constant $A > 0$ such that

$$f(r+s; x, y) \leqq A[f(r; x, y) + f(s; x, y)] \tag{4.5.4}$$

identically in $r > 0$, $s > 0$, $x, y \in R$.

THEOREM 4.5.2. *Any density subordinate to one of special kind is likewise so, with the same A.*

Proof.
$$\begin{aligned} \tilde{f}(u+v; x, y) &= \int_0^\infty f(t; x, y) \, d_t \gamma(t; u+v) \\ &= \int_0^\infty \int_0^\infty f(r+s; x, y) \, d_r \gamma(r; u) \, d_s \gamma(s; v) \\ &\leqq A \int_0^\infty \int_0^\infty f(r; x, y) \, d_r \gamma(r; u) \, d_s \gamma(s; v) \\ &\quad + A \int_0^\infty \int_0^\infty f(s; x, y) \, d_r \gamma(r; u) \, d_s \gamma(s; v) \\ &= A \tilde{f}(u; x, y) + A \tilde{f}(v; x, y). \end{aligned}$$

Note that the property (4.5.4.) is NOT possessed by the Gaussian density in E_1,

$$f(t; x, y) = \frac{1}{t^{\frac{1}{2}}} e^{-(\pi/t)(x-y)^2}, \tag{4.5.5}$$

but it is possessed by the Poisson–Cauchy density

$$\frac{1}{\pi} \frac{t}{(x-y)^2 + t^2}, \tag{4.5.6}$$

for which it was introduced by R.E.A.C. Paley, in its periodic version (2.5.9) on T_1. Also, their characteristic functions are $e^{-\pi t \alpha^2}$ and $e^{-2\pi t|\alpha|}$, so that the second is subordinate to the first.

THEOREM 4.5.3. *For a density of special kind, if we take a Baire function* $t(y)$, $y \in R$, $0 < t < \infty$ *and set up for* $g \geqq 0$ *the transformation*

$$h(x) = \int_R f(t(x); x, y)\, g(y)\, dv_y \equiv \int f(t(x); y, x)\, g(y)\, dv_y, \quad (4.5.7)$$

then we have

$$\int h(x)^2\, dv_x \leqq 4A^2 \int g(y)^2\, dv_y, \quad (4.5.8)$$

Remark. As known, under certain secondary assumptions, by suitably choosing $t(y)$ the function (4.5.7) will be a.e. equal to

$$\bar{h}(x) = \sup_{0<t<\infty} \int_R f(t; x, y)\, g(y)\, dv_y, \quad (4.5.9)$$

in which case (4.5.8) is

$$\int \bar{h}(x)^2\, dv_x \leqq 4A^2 \int g(y)^2\, dy_y, \quad (4.5.10)$$

and this is the manner in which assertions of this sort are usually stated.

Proof. Using Hilbert space theory, if for given $t(x)$ we denote the operator (4.5.7) by $h = Lg$ and introduce the adjoint operator

$$g(y) = L^* \chi \equiv \int f(t(z); z, y)\, \chi(z)\, dv_z,$$

then we have

$$\begin{aligned} L(L^*g) &= \iint f(t(x); y, x) f(t(z); z, y)\, g(z)\, dv_y dv_z \\ &= \int f(t(x) + t(z); x, z)\, g(z)\, dv_z \\ &\leqq A \int f(t(x); x, z)\, g(z)\, dv_z + A \int f(t(z); x, z)\, g(z)\, dv_z \\ &= ALg + AL^*g, \end{aligned}$$

and hence

$$(L^*g, L^*g) \equiv (g, LL^*g) \leqq A(g, Lg) + A(g, L^*g) = 2A(g, L^*g),$$

and this implies $\| Lg \|_2^2 \equiv \| L^*g \|_2^2 \leqq 2A \| g \|_2 \cdot \| L^*g \|_2$
and this is (4.5.8)

4.6. Functions of the Laplace operator

With any given subdivisible process $\{F(t; x_j)\}$ in E_k we associate the semigroup of operators

$$h_t(x_j) = \int_{E_k} g(x_j - y_j)\, d_y F(t; y_j) \equiv L_t g, \tag{4.6.1}$$

and in characteristic functions we write for this also

$$\phi_{h_t}(\alpha_j) = \phi_g(\alpha_j)\, e^{-t\psi(\alpha_j)}. \tag{4.6.2}$$

It follows with the aid of theorem 2.2.4 that these operators are bounded distributive transformations of the Banach spaces L_1 and L_2 into themselves, and we will view them as transformations of the intersection $L_{1,2} \equiv L_1 \cap L_2$ into itself.

THEOREM 4.6.1. *A semigroup of distributive transformations*

$$h_t = L_t g, \tag{4.6.3}$$

$0 \leqq t < \infty$ *of* $L_{1,2}$ *into itself is presentable in the form* (4.6.1) *if and only if* (i) *each* L_t *is commutative with translations and bounded in* L_2*-norm,* (ii) *for* $g \geqq 0$ *we have* $h_t \geqq 0$, *and* (iii) *it converges strongly in* L_2*-norm to identity at the origin,*

$$\lim_{t \downarrow 0} \| L_t g - g \|_2 = 0. \tag{4.6.4}$$

Proof. The 'only if' part is quite easy, and for the proof of the 'if' part the following proposition will be taken as known.

LEMMA 4.6.1. *A bounded linear operator* $h = Lg$ *of* $L_2(E_k)$ *into itself is commutative with translations (if and) only if there is a bounded measurable 'multiplier'* $\chi(\alpha_j)$ *in* E_k *such that*

$$\phi_h(\alpha_j) = \phi_g(\alpha_j)\, \chi(\alpha_j) \quad \text{a.e.} \tag{4.6.5}$$

Now since $L_{1,2}$ is dense in L_2, by assumption (i) of theorem 4.6.1 there exists a family of bounded measurable functions $\{\chi(t; \alpha)\}$ such that

$$\phi_{h_t}(\alpha) = \phi_g(\alpha)\, \chi(t; \alpha), \tag{4.6.6}$$

$$\chi(r; \alpha)\, \chi(s; \alpha) = \chi(r+s; \alpha), \tag{4.6.7}$$

a.e. Next, since $L_{1,2}$ contains $\epsilon^{-\frac{1}{2}k} e^{-\pi\epsilon^{-1}|x|^2}$ it follows by assumption (ii) that, for fixed t, $\chi(t; \alpha)\, e^{-\pi\epsilon|\alpha|^2}$ is a positive-definite function a.e. for each ϵ, and by theorem 3.2.1 the function $\chi(t; \alpha)$ is therefore positive-definite after alteration on a set of measure 0. Finally, assumption (iii) implies

$$\lim_{t \downarrow 0} \int_{E_k} e^{-|\alpha|^2} | 1 - \chi(t; \alpha) |^2\, dv_\alpha \to 0,$$

and all this implies that $\{\chi(t;\alpha)\}$ is a subdivisible process $\{e^{-t\psi(\alpha)}\}$ as claimed.

Next, the following facts from operator theory will be taken as known:

LEMMA 4.6.2. *If $\chi(\alpha_j)$ is a (finite) measurable function in E_k (not bounded) then there is a distributive operator $h=\Delta g$ which is expressed by*

$$\phi_h(\alpha)=\phi_g(\alpha)\,\chi(\alpha). \tag{4.6.8}$$

It is defined for those elements $g \in L_2$ for which

$$\int_{E_k} |\phi_g(\alpha)|^2 (1+|\chi(\alpha)|^2)\,dv_\alpha, \tag{4.6.9}$$

and these elements are dense in L_2; also the operator is self-adjoint if $\chi(\alpha)$ is real-valued, and a (more general) 'normal' operator, if $\chi(\alpha)$ is complex-valued.

These operators are commutative, and if we denote the operators pertaining to $\chi(\alpha)=\alpha_j$, $j=1,\dots,k$, by

$$D_j \equiv -\frac{1}{2\pi i}\frac{\partial}{\partial x_j}, \tag{4.6.10}$$

then the previous operator can be denoted by

$$\Lambda=\chi(D_1,\dots,D_k),$$

according to a known definition of a 'function' of several normal operators which commute.

The application of this is as follows:

THEOREM 4.6.2. *The function* (4.6.1.) *satisfies the 'diffusion equation'*

$$\frac{dh_t}{dt}=-\psi(D_1,\dots,D_k)\,h_t, \tag{4.6.11}$$

the derivative existing in norm in $0 \leq t < \infty$, whenever the initial element $g=h_0$ is one for which $\psi(D_1,\dots,D_k)\,g$ is defined.

In particular, if we introduce the (negative) Laplacean

$$\Delta=-(D_1^2+\dots+D_k^2)=-\frac{1}{4\pi^2}\left(\frac{\partial^2}{\partial x_1^2}+\dots+\frac{\partial^2}{\partial x_k^2}\right), \tag{4.6.12}$$

then for a process subordinate to it, $\psi(\alpha)=\nu(|\alpha|^2)$. we have

$$\frac{dh_t}{dt}=-\nu(\Delta)\,h_t, \tag{4.6.13}$$

whenever $\nu(\Delta)\,g$ exists, at any rate.

Proof. The Plancherel transform of $1/\xi(h_{t+\xi}-h_t)$, $0 \leqq t < \infty$, $0 \leqq t+\xi < \infty$ is

$$\phi_g(\alpha)\, e^{-t\psi(\alpha)} \frac{e^{-\xi\psi(\alpha)}-1}{\xi}$$

and for fixed $t \geqq 0$, this converges in L_2-norm to $\phi_g(\alpha)\, e^{-t\psi(\alpha)}(-\psi(\alpha))$, whenever $\phi_g(\alpha)\,\psi(\alpha) \in L_2$, at any rate.

Turning now to spaces other than E_k we first note as follows:

THEOREM 4.6.3. *The preceding definitions and statements can be adapted to (periodic) functions and subdivisible processes on* T_k.

Now, if a process on T_k consists of densities,

$$f(t;\, x, y) \equiv f(t;\, x-y) \sim \sum_{(m)} e^{-t\psi(m)}\, e^{2\pi i(m,\, x-y)},$$

then on putting $f_m(x) = e^{2\pi i(m,\, x)}$, we may write for this

$$f(t;\, x, y) \sim \sum_{(m)} e^{-t\psi(m)} f_m(x) \overline{f_m(y)}, \tag{4.6.14}$$

and this we now axiomatize as follows. On a measure space (4.4.11) with $v(R)=1$ there is given a countable complete (complex-valued) orthonormal system

$$\int_R f_m(x) \overline{f_{m'}(x)}\, dv_x = \delta_{mm'}, \tag{4.6.15}$$

which is indexed by some labels $\{m\}$, and there is given a stationary Markoff density as previously defined which has an expansion (4.6.14) with exponents for which

$$\operatorname{Re} \psi(m) \geqq 0, \tag{4.6.16}$$

and this expansion is convergent in the following manner. If we take any element $g(x) \in L_2(R)$ and assume that say $g(x) \geqq 0$, and if we introduce the expansion

$$g(x) \sim \sum_{(m)} \gamma_m f_m(x), \tag{4.6.17}$$

then for every $t > 0$ the (non-negative) integral

$$h_t(x) = \int f(t;\, x, y)\, g(y)\, dv_y \equiv L_t g \tag{4.6.18}$$

is rigorously what a formal substitution of (4.6.14) indicates, namely, again an element in $L_2(R)$ and, in fact, the element whose expansion is

$$h_t(x) \sim \sum_{(m)} e^{-t\psi(m)} \gamma_m f_m(x). \tag{4.6.19}$$

Now, with any set of complex numbers $\{\chi(m)\}$ we can associate a normal operator $h=\Lambda g$ on the Hilbert space $L_2(R)$ which is defined for those elements (4.6.17) for which

$$\sum_{(m)} |\gamma(m)|^2 (1+|\chi(m)|^2) < \infty,$$

the value of it being $\Lambda g \sim \sum_{(m)} \chi(m)\gamma(m) f_m(x),$

and we may now state as follows:

THEOREM 4.6.4. *If we are given* (4.6.14) *and* Λ *pertains to the given multipliers* $\{\psi(m)\}$, *and if* g *is an element for which* Λg *is defined, then* (4.6.18) *satisfies*

$$\frac{dh_t}{dt} = -\Lambda h_t, \tag{4.6.20}$$

strongly in $0 \leqq t < \infty$.

Also, if $\tilde{f}(t; x, y)$ *is subordinated to* $f(t; x, y)$ *by* (4.3.2), *then, subject to secondary assumptions we have*

$$\tilde{f}(t; x, y) \sim \sum_{(m)} e^{-t\nu(\psi(m))} f_m(x) \overline{f_m(y)}, \tag{4.6.21}$$

and for the corresponding normal operator we have

$$\tilde{\Lambda} = \nu(\Lambda). \tag{4.6.22}$$

Returning to T_k for a moment, if $\psi(m)$ is real-valued,

$$\psi(m) \geqq 0, \tag{4.6.23}$$

or, equivalently, if the density is symmetric, and if we replace $e^{-2\pi i(m, x)}$ by its real and imaginary parts, then we can also write alternately

$$f(t; x, y) \sim \sum_{(m)} e^{-t\psi(m)} f_m(x) f_m(x), \tag{4.6.24}$$

where the symbols $\{m\}$, $\{f_m(x)\}$ are not quite the same as before and the latter again constitute a complete orthonormal system

$$\int_R f_m(x) f_{m'}(x)\, dv_x = \delta_{mm'} f_m(x), \tag{4.6.25}$$

but now a real-valued one.

In the corresponding analogue to theorem 4.6.4 the density $f(t; x, y)$ is of course symmetric now, and the corresponding operator Λ which now carries again f_m into $\psi(m) f_m$ is at present self-adjoint and not only normal, and this is an important difference indeed. The archtype of such an operator on T_k (or E_k) was the ordinary Laplacean Δ, and with it we formed the functions $\nu(\Delta)$, and these constructions can be undertaken effectively on manifolds in general. In fact, if R is a manifold, compact or not, and $\{g_{ij}\}$ a positive-definite symmetric tensor on it, both of sufficiently high differentiability class C^r, and if we introduce the operator

$$\Delta = -\frac{1}{\sqrt{g}}\frac{\partial}{\partial x^i}\left((\sqrt{g}\,g^{ij}\frac{\partial}{\partial x^j}\right), \tag{4.6.26}$$

and the volume element $dv_x = \sqrt{g}\,dx^1 \ldots dx^k$, then a suitable 'closure' of Δ is self-adjoint, and for all compact R and many noncompact ones there exists a Markoff density for which (4.6.20) is the equation

$$\frac{dh_t}{dt} = -\Delta h_t \tag{4.6.27}$$

now. Also, if R is compact and $v(R) = 1$ then we again have an expansion (4.6.24) with $\{f_m(x)\}$ and $\{\psi(m)\}$ being eigenfunctions and eigenvalues of the operator, the latter occurring multiply, and for many noncompact R analogous expansions exist with the index m not being discrete any longer. Also, in all these cases, the function $f(t;\, x, y)$ itself satisfies the equation

$$\frac{\partial f(t;\, x, y)}{\partial t} = -\Delta_x f(t;\, x, y), \tag{4.6.28}$$

which is formally equivalent to (4.6.27) for 'all' initial functions $h_t = g$, and our subordinate densities $\tilde{f}(t;\, x, y)$ satisfy the corresponding equation

$$\frac{\partial \tilde{f}(t;\, x, y)}{\partial t} = -\nu(\Delta)_x \tilde{f}(t;\, x, y), \tag{4.6.29}$$

in which the right side must be interpreted 'operationally' however, since $\nu(\Delta)$ is not a literal differential expression any more. If we view (4.6.28) as defining a Gaussian process then

$$\frac{\partial \tilde{f}}{\partial t} = -\Delta^{\rho} \tilde{f}, \quad 0 < \rho < 1 \tag{4.6.30}$$

defines the corresponding stable processes say. But all these generalizations are symmetric processes only, and the problem of exhibiting nonsymmetric Markoff densities on manifolds in general remains yet to be broached.

4.7. Multidimensional time variable

The semigroup requirement

$$F(r;\cdot) * F(s;\cdot) = F(r+s;\cdot) \tag{4.7.1}$$

for a subdivisible process

$$\{F(t;\, A)\} \tag{4.7.2}$$

on E_k say, and also for a stationary Markoff density on any (4.4.11), if we ignore continuity, can be set up for elements r, s of any 'semi-group of addition' T_0 in which a commutative associative operation $r+s=t$ is defined. A situation of some interest arises if T_0 is a cone in a Euclidean E_n: (t_ν) according to definition 3.2.1; and if in this case the characteristic function $\chi(t;\alpha_j)$ of (4.7.2) is continuous in (t,α) then it must be of the form

$$\exp[-(t_1\psi_1(\alpha_j)+\ldots+t_n\psi_n(\alpha_j))], \tag{4.7.3}$$

as can be shown. Also, the 'infinitesimal' description of the process is now given by the *system* of differential equations

$$\frac{\partial h_t}{\partial t_\nu} = -\psi_\nu(D_1,\ldots,D_k)\,h_t, \quad \nu=1,\ldots,n, \tag{4.7.4}$$

in which the operators $\psi_\nu(D_1,\ldots,D_k)$ are normal operators which are such that for every $(t_1,\ldots,t_n)$ in T_0 the operator $t_1\psi_1+\ldots+t_n\psi_n$ has no spectrum in the left half-plane.

Assuming that T_0 is a proper cone we take in E_n: (ξ_ν) the cone X whose dual T is the closure of T_0; and in another space E_l: (η_λ) we take a proper cone Y whose dual in E_l: (w_λ) will be denoted by W and, as in section 2.2, we introduce a completely monotone transformation

$$\eta_\lambda = \nu_\lambda(\xi_1,\ldots,\xi_n), \quad \lambda=1,\ldots,l, \tag{4.7.5}$$

from X to Y. This gives rise to Laplace expansions

$$e^{-(w_1\nu_1(\xi)+\ldots+w_l\nu_l(\xi))} = \int_T e^{-(t,\xi)}\,d_t\gamma(t;w),$$

in which $\{\gamma(B;w)\}$ are Lebesgue measures on the Borel field $\{B\}$ of T, and they again have the properties (4.4.2), (4.4.3), (4.4.4) of a subordinator as before.

Obviously, the function

$$\exp\left[-\sum_{\lambda=1}^{l} w_\lambda\nu_\lambda(\psi_1(\alpha),\ldots,\psi_n(\alpha))\right] \tag{4.7.6}$$

describes a process subordinate to (4.7.3), and if (4.7.2) has a density $f(t;x)$ and $\gamma(B;w)$ is a proper subordinator in the sense that it is zero outside T_0, then the distribution function $F(t;x)$ of (4.7.6) has again a density $\tilde{f}(W;x)$ and the relation

$$\tilde{f}(w;x) = \int_{T_0} f(t;x)\,d_t\gamma(t;w)$$

holds.

This again applies to functions on the multitorus and to Markoff densities with expansions

$$\sum_{(m)} \exp[-(t_1\psi_1(m)+\ldots+t_n\psi_n(m))]f_m(x)\overline{f_m(x)}$$

on general spaces as before.

Returning to (4.7.2), if this is Gaussian symmetric for every t in T, then $\psi_\nu(\alpha_j)$ is a symmetric quadratic form $Q_\nu(\alpha_j)$, not necessarily positive semidefinite by itself, and we think that the following ought to be a pertinent definition:

DEFINITION 4.7.1. A subdivisible process $\{\tilde{F}(w; A)\}$ on a proper closed cone W in some E_l: (w_l) is *subordinate to the Gaussian generally*, if for some dimension n, there is given a completely monotone mapping (4.7.5) as described, and a set of quadratic forms $Q_1(\alpha_j), \ldots, Q_n(\alpha_j)$ such that the characteristic function of $\tilde{F}(w; A)$ is of the form

$$\exp\left[-\sum_{\lambda=1}^{l} w_\lambda \nu_\lambda(Q_1(\alpha_j), \ldots, Q_n(\alpha_j))\right].$$

It would be of interest to study this class of processes for any given k and W, say.

4.8. Riemann's functional equation for zeta functions

THEOREM 4.8.1. *In the plane of the complex variable $s=\sigma+it$ let $\chi(s)$ be defined and holomorphic in a domain containing the exterior of a circle*

$$|s| \geqq \rho_0, \tag{4.8.1}$$

and its Laurent decomposition there be

$$\chi(s)=\chi^0(s)+\chi^*(s), \tag{4.8.2}$$

$$\chi^0(s)=\sum_{m=0}^{\infty}\frac{c_n}{s^{n+1}}, \quad \chi^*(s)=\sum_{m=0}^{\infty}\gamma_n s^n, \tag{4.8.3}$$

$$\overline{\lim}\, |c_n|^{1/n} \leqq \rho_0. \tag{4.8.4}$$

If $\chi(s)$ possesses in a right half-plane an absolutely convergent expansion

$$\chi(s)=\int_{-\infty}^{\infty}\phi_r(y)\,e^{-ys}\,dy, \quad \sigma \geqq \sigma_r(>\rho_0) \tag{4.8.5}$$

with a continuous $\phi_r(y)$, and a similar expansion

$$\chi(s)=\int_{-\infty}^{\infty}\phi_l(y)\,e^{-ys}\,dy, \quad \sigma \leqq \sigma_l(<-\rho_0), \tag{4.8.6}$$

in a left half-plane, and if we have

$$\lim_{|t|\to\infty} \chi(\sigma+it)=0 \tag{4.8.7}$$

uniformly in every finite interval $\sigma' \leqq \sigma \leqq \sigma''$, *say, then the difference*

$$\phi_r(y)-\phi_l(y)=p(y) \tag{4.8.8}$$

is an entire function of exponential type, namely,

$$\begin{aligned} p(y) &= \sum_0^\infty \frac{c_n y^n}{n!} \\ &= \frac{1}{2\pi i}\oint_C \chi(s)\, e^{ys}\, ds \equiv \frac{1}{2\pi i}\oint_C \chi^0(s)\, e^{ys}\, ds, \end{aligned} \tag{4.8.9}$$

where C is the circle $|s|=\rho_0$ *say.*

Proof. By Fourier inversion of (4.8.5) we obtain, subject to a criterion,

$$\phi_r(y)\, e^{-y\sigma_r}=\frac{1}{2\pi}\int_{-\infty}^{\infty} \chi(\sigma_r+it)\, e^{yit}\, dt,$$

that is,

$$\phi_r(y)=\frac{1}{2\pi i}\int_{\sigma_r-i\infty}^{\sigma_r+i\infty} \chi(s)\, e^{ys}\, ds, \tag{4.8.10}$$

and similarly

$$\phi_l(y)=\frac{1}{2\pi i}\int_{\sigma_l-i\infty}^{\sigma_l+i\infty} \chi(s)\, e^{ys}\, ds, \tag{4.8.11}$$

and if these integrals are actually convergent, then, on account of (4.8.7), by the use of Cauchy's theorem we obtain for $\phi_r(y)-\phi_l(y)$ the value

$$\frac{1}{2\pi i}\oint_C \chi(s)\, e^{ys}\, ds.$$

However, $\chi^*(s)$ is an entire function and therefore we may herein replace $\chi(s)$ by $\chi^0(s)$, and if we substitute (4.8.3) we obtain the series (4.8.9). In general, the two integrals (4.8.10) and (4.8.11) can be evaluated as limits, for $\epsilon\to 0$, of

$$\frac{1}{2\pi i}\int_{\sigma-i\infty}^{\sigma+i\infty} e^{\epsilon s^2}\chi(s)\, e^{ys}\, ds,$$

(theorem 2.1.3), and if we apply Cauchy's theorem first and put $\epsilon=0$ afterwards, the result follows again.

LEMMA 4.8.1. *For any entire function* (4.8.9), *subject to* (4.8.4), *we have*

$$\int_0^\infty p(y)\, e^{-ys}\, dy=\sum_0^\infty \frac{c_n}{s^{n+1}}\ (\equiv \chi^0(s))$$

for $\sigma > \rho_0$ *and*
$$-\int_{-\infty}^{0} p(y)\, e^{-ys}\, dy = \chi^0(s)$$
for $\sigma < -\rho_0$.

This follows readily by direct substitution of the series (4.8.9) and term-by-term integration.

THEOREM 4.8.2. *If we are given two absolutely convergent integrals*
$$\chi_r(s) = \int_{-\infty}^{\infty} \phi_r(y)\, e^{-ys}\, dy, \quad \chi_l(s) = \int_{-\infty}^{\infty} \phi_l(y)\, e^{-ys}\, dy,$$
the first for $\sigma > \sigma_r$ *and the second for* $\sigma < \sigma_l$, *and if the difference* (4.8.8) *is a function of exponential type, then there exists a holomorphic function* $\chi(s)$ *in a domain* (4.8.1) *with the property* (4.8.7), *such that*
$$\chi_r(s) = \chi(s) \quad \text{for} \quad \sigma > \sigma_r \qquad \textit{and} \qquad \chi_l(s) = \chi(s) \quad \text{for} \quad \sigma < \sigma_l.$$

Proof. It is easily seen that the two integrals in the sum
$$\chi^*(s) = \int_{-\infty}^{0} \phi_r(y)\, e^{-ys}\, dy + \int_{0}^{\infty} \phi_l(y)\, e^{-ys}\, dy$$
are convergent everywhere, thus defining an entire function. By the preceding lemma, and on using $\phi_r(y) = \phi_l(y) + p(y)$, we obtain, for $\sigma > (\sigma_r, \rho_0)$,
$$\chi_r(s) = \int_{-\infty}^{0} \phi_r(y)\, e^{-ys}\, dy + \int_{0}^{\infty} (\phi_l(y) + p(y))\, e^{-ys}\, dy$$
$$= \chi^*(s) + \chi^0(s),$$
and for $\sigma < (\sigma_l, -\rho_0)$ we obtain again
$$\chi_l(s) = \int_{0}^{\infty} \phi_l(y)\, e^{-ys}\, dy + \int_{-\infty}^{0} (\phi_r(y) - p(y))\, e^{-ys}\, dy$$
$$= \chi^*(s) + \chi^0(s).$$
Also, (4.8.7) is easily verified for $\chi^*(s)$, $\chi^0(s)$ separately, q.e.d.

For application we introduce the variable $x = e^{-y}$, $0 < x < \infty$ and write
$$\phi_r(y) = \Phi_r(x), \quad \phi_l(y) = \Phi_l(x), \quad p(y) = P(x),$$
so that we have
$$\Phi_r(x) - \Phi_l(x) = P(x), \tag{4.8.12}$$
where
$$P(x) = \frac{1}{2\pi i} \oint \frac{\chi(s)\, ds}{x^s},$$
and for either $\sigma > \sigma_r$ or $\sigma < \sigma_l$ we have the Mellin inversion formulas so-called
$$\chi(s) = \int_0^{\infty} \Phi(x)\, x^{s-1}\, dx, \quad \Phi(x) = \frac{1}{2\pi i} \int_{\sigma - i\infty}^{\sigma + i\infty} \frac{\chi(s)\, ds}{x^s}.$$

If we have $$\Phi_r(x) = \sum_1^\infty a_n e^{-\lambda_n x}, \quad 0 < \lambda_1 < \lambda_2 < \ldots, \tag{4.8.13}$$
and, for some $\delta > 0$,
$$\Phi_l(x) = \frac{1}{x^\delta} \sum_1^\infty b_n e^{-\mu_n/x}, \quad 0 < \mu_1 < \mu_2 < \ldots, \tag{4.8.14}$$
and $$P(x) = -a_0 + \frac{b_0}{x^\delta},$$
which corresponds to $$\chi^0(s) = -\frac{a_0}{s} + \frac{b_0}{s-\delta},$$
and if we put $\lambda_0 = \mu_0 = 0$, then due to
$$\frac{\Gamma(s)}{\lambda^s} = \int_0^\infty e^{-\lambda x} x^{s-1}\, dx, \quad \sigma > 0,$$
$$\frac{\Gamma(\delta - s)}{\mu^{\delta - s}} = \int_0^\infty \frac{1}{x^\delta} e^{-\mu/x} x^{s-1}\, dx, \quad \delta > \sigma,$$
the following conclusion ensues:

THEOREM 4.8.3. *Given a 'modular relation'*
$$\sum_0^\infty a_n e^{-\lambda_n x} = \frac{1}{x^\delta} \sum_0^\infty b_n e^{-\mu_n/x}, \quad \delta > 0, \tag{4.8.15}$$
if the series $$\xi(s) = \sum_1^\infty \frac{a_n}{\lambda_n^s}, \quad \eta(s) = \sum_1^\infty \frac{b_n}{\mu_n^s} \tag{4.8.16}$$
are absolutely convergent somewhere, then they are meromorphic functions and we have $$\Gamma(s)\,\xi(s) = \Gamma(\delta - s)\,\eta(\delta - s)\ (\equiv \chi(s)), \tag{4.8.17}$$
$$\chi(s) = -\frac{a_0}{s} + \frac{b_0}{s-\delta} + (\text{entire function}). \tag{4.8.18}$$

Conversely, (4.8.16)–(4.8.18) *implies* (4.8.15).

Riemann's own application was to derive from
$$\sum_{-\infty}^\infty e^{-\pi t m^2} = \frac{1}{t^{\frac{1}{2}}} \sum_{-\infty}^\infty e^{-(\pi/t)n^2}$$
the equation $$\Gamma(s)\,\xi(s) = \Gamma(\tfrac{1}{2} - s)\,\xi(\tfrac{1}{2} - s)$$
for the function $\xi(s) = \sum_1^\infty \frac{1}{\pi^s n^{2s}}$. More generally, theorem 2.6.4 implies as follows. If in the notation of that theorem we put
$$\xi(s) = \sum_{(m)}{}' \frac{R_r(m+x)\, e^{2\pi i \Sigma_j (m_j + x_j) y_j}}{\pi^s (Q(m+x))^s},$$
$$\eta(s) = \sum_{(m)}{}' \frac{R_r'(m+y)\, e^{-2\pi i \Sigma_j (m_j + y_j) x_j}}{\pi^s (Q'(m+y))^s},$$

then we have
$$\Gamma(s)\,\xi(s)=\frac{i^r}{(\det Q)^{\frac{1}{2}}}\,\Gamma(r+\tfrac{1}{2}k-s)\,\eta(r+\tfrac{1}{2}k-s),$$

and there are two simple poles at most, and
$$\chi^0(s)=-\frac{\epsilon(x)\,R_r(x)}{s}-\frac{i^r\epsilon(y)\,R_r'(y)}{(\det Q)^{\frac{1}{2}}}\,\frac{1}{r+\frac{1}{2}k-s},$$

where $\epsilon(x)=1$ if $x\equiv 0 \pmod{(m_1,\ldots,m_k)}$ and $\epsilon(x)=0$ otherwise. In particular, for the function
$$\xi(s)=\frac{1}{\pi^s}\Sigma'\,\frac{P_r(m_1,\ldots,m_k)}{(m_1^2+\ldots m_k^2)^s},$$

we have
$$\Gamma(s)\,\xi(s)=i^r\Gamma(r+\tfrac{1}{2}k-s)\,\xi(r+\tfrac{1}{2}k-s),$$

where $P_r(x)$ is any harmonic polynomial of order $r=0,1,\ldots$.

4.9. Summation formulas and Bessel functions in one and several variables

THEOREM 4.9.1. *Formally, if a pair of functions*
$$\{f(\lambda), g(\mu)\} \tag{4.9.1}$$

are connected by the formula
$$g(\mu)=\mu^{-\frac{1}{2}(\delta-1)}\int_0^\infty J_{\delta-1}(2\sqrt{(\mu\lambda)})\,\lambda^{\frac{1}{2}(\delta-1)}f(\lambda)\,d\lambda \tag{4.9.2}$$

in which case the pair
$$\left\{f(\lambda x),\quad \frac{1}{x^\delta}g\left(\frac{\mu}{x}\right)\right\} \tag{4.9.3}$$

is obviously likewise so connected for any $x>0$, then any modular relation
$$\sum_0^\infty a_n e^{-\lambda_n x}=\frac{1}{x^\delta}\sum_0^\infty b_n e^{-\lambda_n/x} \tag{4.9.4}$$

implies the summation formula
$$\sum_0^\infty a_n f(\lambda_n)=\sum_0^\infty b_n g(\mu_n), \tag{4.9.5}$$

and therefore also
$$\sum_0^\infty a_n f(\lambda_n x)=\frac{1}{x^\delta}\sum_0^\infty b_n g\left(\frac{\mu_n}{x}\right). \tag{4.9.6}$$

Proof. We will take as known the formula
$$\frac{1}{x^\delta}e^{-\mu/x}=\mu^{-\frac{1}{2}(\delta-1)}\int_0^\infty J_{\delta-1}(2\sqrt{(\mu\lambda)})\,\lambda^{\frac{1}{2}(\delta-1)}e^{-\lambda x}\,d\lambda, \tag{4.9.7}$$

which simply means that the pair of functions
$$\{e^{-\lambda}, e^{-\mu}\} \tag{4.9.8}$$

is linked by relation (4.9.2). Now, the original modular relation (4.9.4) is a special case of the formula (4.9.6) for this pair of functions, and thus our theorem holds in this special case at any rate. However, starting from (4.9.7), if we multiply (4.9.4) by $d\rho(x)$ and integrate over $(0,\infty)$ then we obtain (4.9.5) for the functions

$$f(\lambda)=\int_0^\infty e^{-\lambda x}\,d\rho(x), \quad g(\mu)=\int_0^\infty \frac{1}{x^\delta}e^{-\mu/x}\,d\rho(x),$$

and if we substitute (4.9.7) we obtain (4.9.2) in this more general case, as claimed.

In this way we have obtained actual criteria, which we will not reproduce here, for the validity of (4.9.5) for completely monotone functions $f(\lambda)$, in particular for $a_n \geqq 0$, $b_n \geqq 0$, but the following alternate procedure was likewise featured. If we extend the Laplace transform (4.9.7) into the complex half-plane $z=x+i\alpha$, $0<x<\infty$, $-\infty<\alpha<\infty$, then by Fourier inversion it becomes equivalent with

$$\frac{1}{2\pi i}\int_{x-i\infty}^{x+i\infty}\frac{e^{-\mu/z+\lambda z}}{z^\delta}\,dz=\begin{cases}\mu^{-\frac{1}{2}(\delta-1)}J_{\delta-1}(2\sqrt{(\mu\lambda)})\,\lambda^{\frac{1}{2}(\delta-1)} & \text{for } \lambda>0\\ 0 & \text{for } \lambda<0,\end{cases} \tag{4.9.9}$$

which is an important formula due to Sonine. And in keeping with this approach, if we start from

$$\sum_0^\infty a_n e^{-\lambda_n(x+i\alpha)}=\frac{1}{(x+i\alpha)^\delta}\sum_0^\infty b_n e^{-\mu_n/(x+i\alpha)}$$

for some $x=\epsilon$ (>0 but small), multiply both sides by $d\sigma(\alpha)$ and integrate over $(-\infty,\infty)$, then this verifies our theorem for appropriate classes of functions representable by integrals of the form

$$f(\lambda)=\int_{-\infty}^\infty e^{-(\epsilon+i\alpha)\lambda}\,d\sigma(\alpha), \quad \int_{-\infty}^\infty |\,d\sigma(\alpha)\,|<\infty,$$

and with some caution we can even let $\epsilon\to 0$ as well. However, we will not reproduce details here, but we will turn to functions in several variables instead, for some first statements at any rate.

In the Euclidean E_k of the points $x=(x_1,\ldots,x_k)$ we take a proper cone which we now denote by P. Its (closed) dual will be denoted by G, its points by $\lambda=(\lambda^1,\ldots,\lambda^k)$ and $\mu=(\mu^1,\ldots,\mu^k)$, and we will also introduce sequences of points $\lambda_n=(\lambda_n^1,\ldots,\lambda_n^k)$ and $\mu_n=(\mu_n^1,\ldots,\mu_n^k)$.

As a generalization of the half-plane $z=x+i\alpha$ we introduce in the space E_{2k} of the k complex variables

$$z_j=x_j+i\alpha_j$$

the point set

$$(x_1, \ldots, x_k) \in P, \quad -\infty < \alpha_j < \infty, \quad j = 1, \ldots, k,$$

and denote it by T_P ('Tube' with basis P). Next, as a generalization of the transformation $y = x^{-1}$ which dominates the structure of (4.9.4), we assume that there is given on P an analytic involution

$$y = Ux, \tag{4.9.10}$$

that is, a transformation $y_j = U_j(x)$ for which

$$UU = \text{identity}, \tag{4.9.11}$$

and we assume that an analytic continuation of (4.9.10) transforms T_P into itself holomorphically. Finally, as a generalization of the denominator x^δ in (4.9.4) we assume given in T_P a holomorphic function $R(z)$ for which $R(U(z)) = (R(z))^{-1}$, $R(z) \neq 0$ in T_P, and $R(x)$ is real for x in P.

As a generalization of (4.9.4) we introduce a relation (if existing) of the form

$$\sum_{n=0}^{\infty} a_n e^{-(\lambda_n, z)} = \frac{1}{R(z)} \sum_{n=0}^{\infty} b_n e^{-(\mu_n, U(z))}, \tag{4.9.12}$$

with λ_n, μ_n in G, and again $(\lambda, z) = \lambda^1 z_1 + \ldots + \lambda^k z_k$, and our aim is to find a suitable transformation

$$g(\mu) = \int_G S(\mu; \lambda) f(\lambda) \, dv_\lambda, \tag{4.9.13}$$

in which $S(\mu; \lambda)$ is meant to generalize the function

$$\mu^{-\frac{1}{2}(\delta-1)} J_{\delta-1}(2\sqrt{(\mu\lambda)}) \lambda^{\frac{1}{2}(\delta-1)},$$

such that (4.9.12) formally implies the summation formula

$$\Sigma a_n f(\lambda_n) = \Sigma b_n g(\mu_n) \tag{4.9.14}$$

for any pair of functions so connected.

For μ in G and λ in E_k we set up the multiple complex integral

$$S(\mu; \lambda) = \frac{1}{(2\pi i)^k} \int_{x_1 - i\infty}^{x_1 - i\infty} \cdots \int_{x_k - i\infty}^{x_k - i\infty} \frac{e^{-(\mu, U(z)) + (\lambda, z)}}{R(z)} \, dz_1 \ldots dz_k \tag{4.9.15}$$

for a point $(x_1, \ldots, x_k)$ in P. If for some fixed $p \geq 1$ we introduce the quantity

$$M_p(x) = \left(\int_{-\infty}^{\infty} \cdots \int_{-\infty}^{\infty} \frac{d\alpha_1 \ldots d\alpha_k}{|R(x_1 + i\alpha_1, \ldots, x_k + i\alpha_k)|^p} \right)^{1/p},$$

then the integral (4.9.15) obviously exists if $M_1(x)<\infty$, and if, more precisely, we have

$$\sup_{x\in A}|M_1(x)|<\infty \tag{4.9.16}$$

for every compact subset A of P then the integral is indeed independent of the point x chosen, as can be proven by shifting the coordinates x_j individually, and altogether several times. Furthermore, if also $M_2(x)<\infty$, then by Plancherel equation we have

$$\int_{E_k}|S(\mu;\lambda)|^2 e^{-2(\lambda,x)}\,dv_\lambda=(2\pi)^{-k}\int_{E_k}\frac{e^{-2(\mu,\,\mathrm{Re}\,U(z))}}{|R(x+i\alpha)|^2}\,dv_\alpha, \tag{4.9.17}$$

and if we assume that for every x^0 in P we have

$$\sup_{1\leqq t<\infty} M_2(tx^0)<\infty, \tag{4.9.18}$$

then, because of $(\mu,\mathrm{Re}\,U(z))\geqq 0$, (4.9.17) implies that we have $S(\mu;\lambda)=0$ for λ not in G.

THEOREM 4.9.2. *Under the assumptions* (4.9.16) *and* (4.9.18) *we have for μ in G,*

$$\frac{1}{(2\pi i)^k}\int_{x_1-i\infty}^{x_1+i\infty}\cdots\int_{x_k-i\infty}^{x_k+i\infty}\frac{e^{-(\mu,U(z))+(\lambda,z)}}{R(z)}\,dz_1\ldots dz_k =\begin{cases}S(\mu;\lambda) & \text{for } \lambda \text{ in } G,\\ 0 & \text{for } \lambda \text{ not in } G,\end{cases} \tag{4.9.19}$$

the integral being independent of x in P.

For $R(x)=x^\delta$ the two assumptions are fulfilled only for $\delta>1$ and $\delta>\frac{1}{2}$ respectively, whereas the conclusion happens also to hold for $\delta>0$. Now, it could be shown that our theorem also holds if the assumptions are fulfilled for $M_p(x)$ for some $p>1$ instead of only $p=1,2$, and such a generalization would include $\delta>0$ in particular.

Next, by partial differentiation under the integral sign in (4.9.19) the following conclusion can be obtained:

THEOREM 4.9.3. *If $p(z_1,\ldots,z_k)$, $q(z_1,\ldots,z_k)$ are polynomials for which*

$$q(U(z)).p(z)=1, \tag{4.9.20}$$

and if $q^*(z)$ is the polynomial adjoint to $q(z)$, then for any $R(z)$ we have the two-fold partial differential equation

$$q^*\left(\frac{\partial}{\partial\mu_j}\right)p\left(\frac{\partial}{\partial\lambda_j}\right)S(\mu;\lambda)=S(\mu;\lambda), \tag{4.9.21}$$

subject to convergence assumptions which bear on $R(z)$, this equation being usually of an elliptic-hyperbolic type peculiarly 'mixed'.

For the ordinary Bessel function we obtain in particular

$$\frac{\partial^2}{\partial\mu\partial\lambda}(\mu^{-\frac{1}{2}\gamma}J_\gamma(2\sqrt{(\mu\lambda)}\,\lambda^{\frac{1}{2}\gamma}) = -\mu^{-\frac{1}{2}\gamma}J_\gamma(2\sqrt{(\mu\lambda)})\,\lambda^{\frac{1}{2}\gamma}, \quad (4.9.22)$$

which is equivalent to the classical equation, of course.

Frequently we will have $q(z)=p(z)$, $R(z)=p(z)^\delta$, and then in addition to the 'principal' cone P there may be other cones to which our analysis applies, thus giving rise to 'Bessel' functions of 'second' kind, and possibly also of 'third' kind, etc.

By Fourier inversion of (4.9.19) we obtain as follows:

THEOREM 4.9.4. *For z in T_P we have*

$$\frac{e^{-(\mu,U(z))}}{R(z)} = \int_G S(\mu;\lambda)\,e^{-(\lambda,z)}\,dv_\lambda \quad (4.9.23)$$

(the integral converging absolutely), and after replacing z by $U(z)$ we also have the inverse relation

$$e^{-(\mu,z)} = \int_G S(\mu;\lambda)\frac{e^{-(\lambda,U(z))}}{R(z)}\,dv_\lambda. \quad (4.9.24)$$

Now for a finite sum $\quad f(\lambda) = \sum_\rho c_\rho\,e^{-(\lambda,z^\rho)} \quad (4.9.25)$

the transform (4.9.13) has the value

$$g(\mu) = \sum_\rho c_\rho \frac{e^{-(\mu,U(z^\rho))}}{R(z)},$$

and therefore theorem 4.9.4 leads to the following conclusion:

THEOREM 4.9.5. *Formally,* (4.9.14) *holds and the transformation* (4.9.13) *is self-inversive, meaning that it implies*

$$f(\mu) = \int_G S(\mu;\lambda)\,g(\lambda)\,dv_\lambda. \quad (4.9.26)$$

An obvious (but important) multidimensional case arises if P is the octant $x_1>0,\ldots,x_k>0$ and $R(x)$ is a product $x_1^{\delta_1}\ldots x_k^{\delta_k}$ for positive exponents $\delta_1,\ldots,\delta_k$, equal or not.

Another situation arises (and apart from variants and combinations it is the only nonobvious one known) if for a dimension $k=m(m+1)/2$ we consider the symmetric real matrices

$$X \equiv \{x_{pq}\}_{p,q=1,\ldots,m},$$

and as independent variables take the quantities x_{pp}, $1\leqq p\leqq m$, $\sqrt{2}\,x_{pq}$, $1\leqq p<q\leqq m$. The domain P shall consist of matrices which are strictly positive-definite, and the dual space consists then of matrices

which are positive-semidefinite, the inner product (λ, x) being then the matrix trace,

$$\operatorname{trace}(\lambda x) \equiv \sum_{p,q=1}^{m} \lambda^{pq} x_{pq}.$$

The involution $U(z)$ is the matrix reciprocation z^{-1}, and $R(z)$ is $(\det(z))^{\delta}$, for δ sufficiently high. The function $S(\mu; \lambda)$ is then the multiple integral

$$\frac{1}{(2\pi i)^k} \int \cdots \int \frac{e^{-\operatorname{trace}(\mu z^{-1}) + \operatorname{trace}(\lambda z)}}{(\det(z))^{\delta}} \, dz_1 \ldots dz_k,$$

and for this function we showed, in a somewhat more general setting, that the product

$$I(\mu; \lambda) = (\det | \mu \lambda^{-1} |)^{\delta - \frac{1}{2}(m+1)} S(\mu; \lambda),$$

which is the actual generalization of $J_{\delta-1}(2\sqrt{(\mu\lambda)})$, is symmetric in μ, λ,

$$I(\mu; \lambda) = I(\lambda; \mu),$$

and this is a rather nontrivial fact, apparently.

CHAPTER 5

STOCHASTIC PROCESSES AND CHARACTERISTIC FUNCTIONALS

5.1. Directed sets of probability spaces

We are recalling the definitions in section 3.7 to which we will refer soon.

DEFINITION 5.1.1. A set of elements Λ: (λ) is called *directed* if for some pairs of elements there is given an order relation '$<$' such that (1) $\lambda < \lambda$, (2) $\lambda < \mu$ and $\mu < \nu$ implies $\lambda < \nu$, (3) $\lambda < \mu$ and $\mu < \lambda$ implies $\lambda = \mu$ and, what is important, (4) given λ, μ there is an element ν such that $\lambda < \nu$, $\mu < \nu$ simultaneously.

DEFINITION 5.1.2. If we are given a family of sets

$$\{\Omega_\lambda: (\omega_\lambda)\} \tag{5.1.1}$$

which is indexed by a directed set, and if for every λ, μ with $\lambda < \mu$ there is given a transformation

$$\omega_\lambda = f_{\lambda\mu}(\omega_\mu) \tag{5.1.2}$$

from Ω_μ to Ω_λ, also called *projection*, such that

$$f_{\lambda\lambda} = \text{identity} \tag{5.1.3}$$

and

$$f_{\lambda\nu} = f_{\lambda\mu}(f_{\mu\nu}) \quad \text{for} \quad \lambda < \mu < \nu; \tag{5.1.4}$$

then the so-called *projective limit* of the family—which we will denote by Ω_∞: (ω_∞) or also by Ω: (ω)—is defined as follows. A point ω_∞ is any assemblage of points

$$\omega_\infty = \{\omega_\lambda\} \tag{5.1.5}$$

from the sets (5.1.1) such that for each λ in Λ there occurs one point ω_λ from Ω_λ, also called the λth *component* of ω_∞ and that for any $\lambda < \mu$ these components are connected by (5.1.2). We are also introducing the mapping

$$\omega_\lambda = f_{\lambda\infty}(\omega_\infty) \tag{5.1.6}$$

to the λth component for which we obviously have

$$f_{\lambda\infty} = f_{\lambda\mu} f_{\mu\infty}, \quad \lambda < \mu, \tag{5.1.7}$$

and we call it the *projection* from Ω_∞ to Ω_λ.

Also, we call our family (5.1.1) *simply maximal* if all projections $f_{\lambda\mu}, f_{\lambda\infty}$ are into all of Ω_λ, so that in particular every point ω_λ in Ω_λ, for every λ, is the λth component of at least one point in Ω_∞.

We will call it *sequentially maximal* if for every increasing sequence of indices

$$\lambda_1 < \lambda_2 < \lambda_3 < \ldots \tag{5.1.8}$$

(which may be finite) and any choice of points $\omega^0_{\lambda_r}$ in Ω_{λ_r} with

$$\omega^0_{\lambda_r} = f_{\lambda_r \lambda_s}(\omega^0_{\lambda_s}), \quad r > s, \tag{5.1.9}$$

there is point ω^0 in Ω_∞ for which $\omega^0_{\lambda_r} = f_{\lambda_r \infty}(\omega^0)$, $r = 1, 2, \ldots$.

DEFINITION 5.1.3. A *stochastic family* is a family of *probability spaces*

$$\{\Omega_\lambda;\ \mathscr{S}_\lambda;\ P_\lambda\} \tag{5.1.10}$$

which is indexed by a directed set together with *simply maximal* transformations (5.1.2) from Ω_μ to Ω_λ such that each $f_{\lambda\mu}$ is also a *consistent* mapping of the entire probability space

$$\{\Omega_\mu;\ \mathscr{S}_\mu;\ P_\mu\} \tag{5.1.11}$$

into the entire (5.1.10).

We call the stochastic family *topological* if each probability space (5.1.10) is topological and the mappings (5.1.2) are continuous.

Now, due to the (simple) maximality stipulated, each $f_{\lambda\infty}$ generates a probability space

$$\{\Omega_\infty;\ \mathscr{S}^*_\lambda;\ P^*_\lambda\} \tag{5.1.12}$$

such that

$$\mathscr{S}^*_\lambda = f^{-1}_{\lambda\infty}(\mathscr{S}_\lambda), \quad P^*_\lambda(S^*_\lambda) = P_\lambda(f_{\lambda\infty}(S^*_\lambda)), \tag{5.1.13}$$

and we denote by $\mathscr{S}^* \equiv \bigcup_\lambda \mathscr{S}^*_\lambda$ the union of all sets in all σ-algebras $\mathscr{S}^*_\lambda$, this being a finitely additive algebra of sets in Ω_∞, obviously. If a set S^* occurs in both $\mathscr{S}^*_\lambda$ and $\mathscr{S}^*_\mu$, and if $\lambda < \mu$ then our consistency assumptions imply

$$P^*_\lambda(S^*) = P^*_\mu(S^*); \tag{5.1.14}$$

and for arbitrary indices λ, μ we can find a third ν, so that $\lambda < \nu$ and $\mu < \nu$ and thus (5.1.14) holds in all cases. If therefore we denote the common value of the two sides in (5.1.14) by $P^*(S^*)$, then we are led to introduce a 'finitely additive' probability space

$$\{\Omega;\ \mathscr{S}^*;\ P^*\}, \tag{5.1.15}$$

in which, however, $\mathscr{S}^*$ and P^* are only finitely additive both.

DEFINITION 5.1.4. We say that a stochastic family has the *Kolmogoroff property* if there is a σ-extension P of the measure P^* from $\mathscr{S}^*$ to its σ-closure $\mathscr{S}$; and we then also call the probability space

$$\{\Omega;\ \mathscr{S}; P\}, \tag{5.1.16}$$

the projective limit of the family (5.1.10) and the entire set-up a *stochastic process*.

THEOREM 5.1.1. *If a stochastic family is topological and sequentially maximal then it has the Kolmogoroff property, and thus gives rise to a stochastic process.*

Proof. By a general measure-theoretic criterion of Kolmogoroff it suffices to show that in the given circumstances there exists no infinite sequence of element S_r^* in $\mathscr{S}^*$ for which we have

$$S_1^* \supset S_2^* \supset S_3^* \supset \ldots \to 0 \tag{5.1.17}$$

and

$$P^*(S_r^*) \geqq \alpha > 0, \quad r = 1, 2, \ldots, \tag{5.1.18}$$

simultaneously. Suppose that such a sequence does exist. We can then find a sequence of indices (5.1.8) such that $S_r^* \in \mathscr{S}_{\lambda_r}^*$, and if we form $S_r = f_{\lambda_r \infty}(S_r^*)$ then there is a compact set $\tilde{C}_r$ in Ω_{λ_r} such that

$$\tilde{C}_r \subset S_r, \quad P_{\lambda_r}(\tilde{C}_r) \geqq P_{\lambda_r}(S_r) - \frac{\alpha}{4^r}$$

for all r. We put $\tilde{C}_r^* = f_{\lambda_r \infty}^{-1}(\tilde{C}_r)$, and then

$$C_r^* = \prod_{\rho=1}^{r} \tilde{C}_\rho^* \equiv \tilde{C}_r^* - \sum_{\rho=1}^{r-1} (S_\rho^* - \tilde{C}_\rho^*),$$

so that obviously $\quad C_r^* \subset S_r^*, \quad C_0^* \supset C_2^* \supset \ldots \to 0, \qquad (5.1.19)$

$$P^*(C_r^*) \geqq P^*(\tilde{C}_r^*) - \alpha\left(\frac{1}{4} + \ldots + \frac{1}{4^{r-1}}\right) \geqq \tfrac{1}{2}\alpha > 0,$$

and if we form back the sets $C_r = f_{\lambda_r \infty}(C_r^*)$ in Ω_{λ_r} then they have the following properties. Each C_r is compact; we have $P_{\lambda_r}(C_r) = P^*(C_r^*) > 0$, so that C_r is nonempty; and we have $f_{\lambda_r \lambda_s}(C_s) \subset C_r$ for $r < s$. On the basis of this we can pick a point in each C_r, which we denote by ω_r^r, and to this we add the further points $\omega_r^s = f_{\lambda_r \lambda_s}(\omega_s^s)$, $r < s$, and we can assert that the sequence $\omega_r^r, \omega_r^{r+1}, \ldots$, has a limiting point in C_r. Next, if we replace $\{\lambda_r\}$ by a subsequence of itself we may even assume that the limits

$$\lim_{s \to \infty} \omega_r^s = \omega_{\lambda_r}^0 \in C_r$$

exist, and from the continuity of $f_{\lambda\mu}$ it follows that (5.1.9) likewise holds. The sequential maximality now implies that there is a point ω in Ω with these limits as components, and this means that the sets C_r^* have a point in common, which contradicts (5.1.19). Hence the theorem.

We do not know whether the sequential maximality is really needed, and also whether the limit space (5.1.16) is likewise topological, and if the given spaces (5.1.10) are strictly topological to what extent the space (5.1.16) is likewise so.

A measurable function $F^0_\lambda(\omega_\lambda)$ on (5.1.10) gives rise to a function

$$F_\lambda(\omega) = F^0_\lambda(f_{\lambda\infty}(\omega_\lambda))$$

on (5.1.16), again measurable, and as previously stated we also have

$$\int_{\Omega_\lambda} F^0_\lambda(\omega_\lambda)\, dP_\lambda = \int_\Omega F_\lambda(\omega)\, dP \quad (\equiv E\{F_\lambda\}). \tag{5.1.20}$$

An interesting situation arises if we are given a monotonely increasing family of measurable functions $\{F_\lambda(\omega)\}$ for which we have

$$0 \leqq F_\lambda(\omega) \leqq F_\mu(\omega), \quad \lambda \leqq \mu. \tag{5.1.21}$$

If the index set Λ is not countable then the classical theorem that the limit of the integral is the integral of the limit cannot be asserted literally, but, for instance, the following version of it remains:

THEOREM 5.1.2. *Given measurable functions with property* (5.1.21), *if we admit limit values* $+\infty$ *then there exists a measurable function* $F(\omega)$ *as follows. For each increasing sequence of indices* $\{\lambda_r\}$ *we have*

$$\lim_{r\to\infty} F_{\lambda_r}(\omega) \leqq F(\omega), \quad \text{a.e.,} \tag{5.1.22}$$

and there is some such sequence for which equality holds; so that if any other $F^*(\omega)$ *satisfies all relation* (5.1.22), *then* $F(\omega) \leqq F^*(\omega)$, a.e.

Proof. Assume first that we also have

$$0 \leqq F_\lambda(\omega) \leqq 1 \tag{5.1.23}$$

for all λ, ω, choose an increasing sequence $\{\lambda_r\}$ for which

$$\sup_r E\{F_{\lambda_r}\} = \sup_\lambda E\{F_\lambda\}, \tag{5.1.24}$$

and put

$$F(\omega) = \sup_r F_{\lambda_r}(\omega).$$

If for any other increasing sequence $\{\mu_s\}$ we put

$$G(\omega) = \sup_s F_{\mu_s}(\omega), \tag{5.1.25}$$

then we can find a third increasing sequence $\{\nu_t\}$ such that for every pair r, s there exists a t for which $\lambda_r \leqq \nu_t$, $\mu_s \leqq \nu_t$. For the function $H(\omega) = \sup_t F_{\nu_t}(\omega)$ we thus have $F(\omega) \leqq H(\omega)$, $G(\omega) \leqq H(\omega)$, and in particular $E\{F\} \leqq E\{H\}$. A comparison with (5.1.24) now implies $E\{F\} = E\{H\}$, and hence $H(\omega) = F(\omega)$ a.e. Therefore, we also have $G(\omega) \leqq F(\omega)$, a.e. for every function (5.1.25), and this proves our theorem for (5.1.23).

If this additional assumption is not given, then we can introduce the new functions

$$F^*_\lambda(\omega) = 1 - e^{-uF_\lambda(\omega)} \tag{5.1.26}$$

for some fixed $u > 0$ for which assumptions (5.1.21) and (5.1.23) hold both, and the theorem follows.

We have introduced the transformation (5.1.26) rather than another one because of the following theorem which will be suitable for applications:

THEOREM 5.1.3. *If all functions are as in Theorem* 5.1.2 *and if we introduce the functions*

$$\phi_\lambda(u)=\int_\Omega e^{-uF_\lambda(\omega)}\,dP(\omega), \quad \phi(u)=\int_\Omega e^{-uF(\omega)}\,dP(\omega), \tag{5.1.26}$$

then obviously

$$\phi(u)=\inf_\lambda \phi_\lambda(u), \tag{5.1.27}$$

and also

$$\phi(0+)=P\{F(\omega)<\infty\}, \tag{5.1.28}$$

that is, $\phi(+0)=P(S^0)$, *where* S^0 *is the set where* $F(\omega)<\infty$. *Thus,* $F(\omega)$ *is finite almost everywhere if and only if* $\Phi(0+)=1$, *and it is infinite almost everywhere if and only if* $\Phi(u)=0$ *for some and then for all* $u>0$.

In fact,

$$\phi(u)=\int_{S^0} e^{-uF(\omega)}\,dP(\omega),$$

and for $u\downarrow 0$ this converges increasingly to

$$\int_{S^0} dP(\omega)=P(S^0).$$

Also, since $\phi(u)$ is completely monotone in $0<u<\infty$ it is either everywhere >0 or $\equiv 0$, and hence the last statement in the theorem.

5.2. Markoff processes

If T is any set and $\{\Delta\}$ a family of its subsets which is closed under set addition $\Delta_1 \cup \Delta_2$ then the point set inclusion $\Delta' \subset \Delta''$ defines an order relation by which $\{\Delta\}$ becomes a directed set suitable for indexing. In our first application T will be the half-line $0<t<\infty$, and the subsets will be the finite sets

$$\lambda=(t^1, \ldots, t^l) \tag{5.2.1}$$

whose totality is obviously so closed.

We take, as in section 3.7, a measure space

$$\{R; \mathscr{B}; v\} \tag{5.2.2}$$

and denote by

$$\{R^l; \mathscr{B}^l; v^l\} \tag{5.2.3}$$

the familiar l-fold product of (5.2.2) with itself, so that in particular

$$R^l=R\times R\times \ldots \times R, \quad l \text{ times}, \tag{5.2.4}$$

and we also fix a point x^0 of R, put $t^0=0$, and order the points in (5.2.1) thus:

$$(0=t^0<)\, t^1<t^2<\ldots<t^l. \tag{5.2.5}$$

With any Markoff chain density $f(r, x; s, y)$ on (5.2.2) we now set up the non-negative set function

$$F^l(B^l; x^0) = \int_{B^l} \left[\prod_{p=1}^{l} f(t^{p-1}, x^{p-1}; t^p, x^p) \right] dv^l(x)$$

$$= \int_{B^l} \prod_{p=1}^{l} [f(t^{p-1}, x^{p-1}; t^p, x^p)\, dv(x^p)], \qquad (5.2.6)$$

and if for $B^l = R^l$ we evaluate the integral by integrating successively with respect to $x^l, x^{l-1}, \ldots, x^2, x^1$ in this order, then the property

$$\int_R f(t^{p-1}, x^{p-1}; t^p, x^p)\, dv(x^p) = 1 \qquad (5.2.7)$$

implies $F^l(R^l; x^0) = 1$. Therefore

$$\Omega_\lambda \equiv R^l; \quad \mathscr{S}_\lambda \equiv \mathscr{B}^l; \quad P_\lambda(B^l) \equiv F^l(B^l; x^0) \qquad (5.2.8)$$

is a probability space, and we associate it with the index (5.2.1) subject to (5.2.5). Given an index $\mu > \lambda$, we denote it by

$$\mu = (t^1, \ldots, t^l; t^{l+1}, \ldots, t^m),$$

and we must permit the new points $t^{l+1}, \ldots, t^m$, if any, to be distributed on the line in any manner whatsoever. We now put

$$\Omega_\mu = R^m; \ \mathscr{S}_\mu \equiv B^m; \quad P_\mu(B^m) \equiv F^m(B^m; x^0), \qquad (5.2.9)$$

denote the points of Ω_λ by $(x^1, \ldots, x^l)$ and those of Ω_μ by

$$(y^1, \ldots, y^l; y^{l+1}, \ldots, y^m),$$

and introduce the 'literal' projection

$$x^p = y^p; \quad p = 1, \ldots, l, \qquad (5.2.10)$$

from Ω_μ to Ω_λ, denoting it $\omega_\lambda = f_{\lambda\mu}(\omega_\mu)$, and we claim that our family (5.2.8) is made into a stochastic family hereby. In fact, properties (5.1.3), (5.1.4) and $f_{\lambda\mu}^{-1}(\mathscr{S}_\lambda) \subset \mathscr{S}_\mu$ are obvious, and for the proof of the last and decisive property

$$F^l(B^l; x^0) = F^m(f_{\lambda\mu}^{-1}(B^l); x^0), \qquad (5.2.11)$$

it is enough to deal with the case $m = l+1$ only, as is easily seen. We denote the additional point $t^{l+1} \equiv t^m$ by s and we have either (i) $t^{p-1} < s < t^p$ for some $p = 1, \ldots, l$ or (ii) $t^l < s$. In either case the set $f_{\lambda u}^{-1}(B^l)$ is $B^l \times R$, and in case (i) say, a point of the set may be denoted by $(x^1, \ldots, x^{p-1}, x, x^p, \ldots, x^l)$ accordingly, and the expression (5.2.6) for $F^{l+1}(B^{l+1}; x^0)$ is then

$$\int_{B^l \times R} \ldots f(t^{p-1}, x^{p-1}; s, x) f(s, x; t^p, x^p)\, dv(x) \ldots,$$

where the dots indicate factors not involving (s, x). By (4.4.13) however this is

$$\int_{B^l} \ldots f(t^{p-1}, x^{p-1}; t^p, x^p) \ldots,$$

which is the original expression for $F^l(B^l; x^0)$ itself, as claimed. In case (ii) relation (5.2.7) must be used for $p=l+1$.

Finally, the projections (5.2.10) are sequentially maximal, and finite products of strictly topological spaces are again so, and hence the following definition and theorem:

DEFINITION 5.2.1. The stochastic family (5.2.8) with projections (5.2.10) will be called a *Markoff process* (for the initial point x^0) if the Kolmogoroff property is present.

THEOREM 5.2.1. *If the underlying measure space* (5.2.2) *is strictly topological then the Kolmogoroff property is present, and we have indeed a Markoff process.*

If the process (that is, the underlying density) is space homogeneous then to a certain extent it is the same process for all initial points x^0, as the following statements will imply in which more generally than before we put

$$0 \leqq t^0 < t^1 < t^2 < \ldots < t^l,$$

where not necessarily $t^0 = 0$.

THEOREM 5.2.2. *If the process is homogeneous for a transitive set of transformations* $\{U\}$ *of* (5.2.2), *and if a bounded Baire function* $\phi(x^0, x^1, \ldots, x^l)$ *on* R^{l+1} *has the invariance property*

$$\phi(Ux^0, Ux^1, \ldots, Ux^l) = \phi(x^0, x^1, \ldots, x^l), \tag{5.2.12}$$

then there is a function $A_\phi(t^0, t^1, \ldots, t^l)$ *such that*

$$\int_{R^l} \phi(x^0, x^1, \ldots, x^l)\, dF^l(x^p; x^0) = A_\phi(t^0, t^1, \ldots, t^l) \tag{5.2.13}$$

for all x^0, *and if we have in particular*

$$\phi(x^0, x^1, \ldots, x^l) \equiv \prod_{p=1}^{l} \phi_p(x^{p-1}, x^p) \tag{5.2.14}$$

$$\phi_p(Ux, Uy) = \phi_p(x, y), \quad p = 1, \ldots, l \tag{5.2.15}$$

then $$A_\phi(t^0, t^1, \ldots, t^l) = A_{\phi_1}(t^0, t^1)\, A_{\phi_2}(t^1, t^2) \ldots A_{\phi_l}(t^{l-1}, t^l). \tag{5.2.16}$$

Proof. By (4.4.18) we have $F^l(UB; Ux^0) = F^l(B; x^0)$, and if we denote the integral (5.2.13) by $A_\phi(x^0; t^p)$, then the substitution $x^p \to U(x^p)$, $p = 0, 1, 2, \ldots, l$, leads to

$$A_\phi(x^0; t^p) = A_\phi(Ux^0; t^p),$$

and since a transitive group carries any point x^0 into any other point y^0 this is indeed independent of x^0. Therefore, we have

$$\int_R \phi_p(x,y) f(r,x;\, s,y)\, dv(y) = A_{\phi_p}(r,s),$$

and (5.2.16) follows if we evaluate (5.2.13) by integrating with respect to $x^l, x^{l-1}, \ldots, x^1$ in this order.

If (5.2.2) is the Euclidean

$$\{E_k;\, \mathscr{A}_k;\, v_k\} \tag{5.2.17}$$

with translations, for any $k \geqq 1$, and if in $(E_k)^l \equiv R^l$: $(x^1, \ldots, x^l)$ we replace, for fixed x^0, the (vectorial) components $x^1, \ldots, x^l$ by the new ones

$$\xi^1 = x^1 - x^0, \quad \xi^2 = x^2 - x^1, \quad \ldots, \quad \xi^l = x^l - x^{l-1}, \tag{5.2.18}$$

then space homogeneity means that we have

$$f(r,x;\, s,y) = f(r,s;\, y-x), \tag{5.2.19}$$

$$\phi(x^0, x^1, \ldots, x^l) = \phi(\xi^1, \ldots, \xi^l), \quad \phi_p(x^{p-1}, x^p) = \phi_p(\xi^p),$$

and because of $dv_k(x^1) \ldots dv_k(x^l) \equiv dv_k(\xi^1) \ldots dv_k(\xi^l)$, we have

$$A_\phi(t^0, t^1, \ldots, t^l) = \int_{(E_k)^l} \phi(\xi^1, \ldots, \xi^l)\, d_\xi F^l(\xi), \tag{5.2.20}$$

where

$$F^l(B^l) = \int_{B^l} \prod_{p=1}^{l} f(t^{p-1}, t^p;\, \xi^p)\, dv_k(\xi^p). \tag{5.2.21}$$

This gives rise to several remarks.

Remark 1. If we associate the entity ξ^p with the interval $t^{p-1} < t \leqq t^p$, then we obtain an additive *interval* function and the probability measures (5.2.21) lead to a stochastic family and process for such, whereas the Markoff process itself pertains to *path* and *point* functions as such. The 'randomization' of interval functions will be presented later in a more comprehensive context (section 5.5).

Remark 2. We note, however, immediately that the 'density' (5.2.19) can be replaced by a more general 'distribution' $F(r,s;\, A_\xi)$ with the properties $F(r,s;\, A) \geqq 0$, $F(r,s;\, E_k) = 1$ and

$$F(r,s;\cdot) * F(s,t;\cdot) = F(r,t;\cdot)$$

for $r < s < t$, not only for a generalization of (5.2.21) which is

$$\int_{E_k} \cdots \int_{E_k} \prod_{p=1}^{l-1} F(t^{p-1}, t^p;\, dA_{\xi_p})\, F(t^{l-1}, t^l;\, A_{\xi l}),$$

but also for a generalization of (5.2.6) which is

$$\int_{B^l} \prod_{j=1}^{l-1} F(t^{p-1}, t^p;\, dB_{x^p} - x^{p-1})\, F(t^{l-1}, t^l;\, B_{x^l} - x^{l-1}),$$

and we have then in particular

$$A_{\phi_p}(t^{p-1}, t^p) = \int_{E_k} \phi_p(\xi)\, d_\xi F(t^{p-1}, t^p;\, \xi).$$

Thus, if our Euclidean chain is also stationary, $F(r, s;\, A) = F(s-r;\, A)$, then in forming the integral

$$A_{\phi_p}(t^{p-1}, t^p) \equiv A_{\phi_p}(t^p - t^{p-1}) = \int_{E_k} \phi_p(\xi)\, d_\xi F(t^p - t^{p-1};\, \xi)$$

we will be able to employ the most general subdivisible process $\{F(r;\, A)\}$ as previously analyzed. All this applies of course also to locally compact Abelian groups other than E_k.

Remark 3. Next, theorem 5.2.1 applies to (5.2.17), so that a limiting space exists and we claim that for a density (5.2.19) the quantities (5.2.18) are stochastically independent in the sense that

$$E\{\phi_1(\xi^1) \dots \phi_l(\xi^l)\} = \prod_{p=1}^{l} E\{\phi_p(\xi^p)\}. \tag{5.2.22}$$

In fact, as is easily verified, this is precisely the meaning of the relation (5.2.16) now.

THEOREM 5.2.3. *Conversely, if we are given a Markoff density $f(r, x;\, s, y)$ on* (5.2.17) *which is continuous in (x, y) and, what is restrictive, satisfies*

$$f(0, x^0;\, r, y) > 0, \tag{5.2.23}$$

and not only $\geqq 0$, for all $r > 0$ and all y in E_k and x^0 fixed, and if for all $0 < r < s$, the two vectors

$$\xi = x(r) - x^0, \quad \eta = x(s) - x(r)$$

are stochastically independent on (5.1.16) *then $f(r, x;\, s, y)$ depends on $r, s, y-x$ only.*

Proof. Put $x^0 = 0$, say. We then have

$$E\{\phi(\xi)\, \psi(\eta)\} = \iint \phi(\xi)\, \psi(\eta) f(0, 0;\, r, \xi) f(r, \xi;\, s, \xi+\eta)\, dv(\xi)\, dv(\eta),$$

$$E\{\phi(\xi)\} = E\{\phi(\xi) \cdot 1\} = \int \phi(\xi) f(0, 0;\, r, \xi)\, dv(\xi)$$

$$E\{\psi(\eta)\} = E\{1 \cdot \psi(\eta)\} = \int \psi(\eta) f(r, s;\, \eta)\, dv(\eta)$$

where $$f(r, s;\, \eta) \equiv \int f(0, 0;\, r, \zeta) f(r, \zeta;\, s, \zeta+\eta)\, dv(\zeta).$$

Now, if we are to have

$$E\{\phi(\xi)\,\psi(\eta)\}=E\{\phi(\xi)\}\,E\{\psi(\eta)\}$$

for all bounded Baire functions $\phi(\xi), \psi(\eta)$, then this implies

$$f(0,0;\,r,\xi)f(r,\xi;\,s,\xi+\eta)=f(0,0;\,r,\xi)f(r,s;\,\eta),$$

except for a set of measure 0 in (ξ,η), the set depending on r,s. However, by our special assumption (5.2.23) we can divide through by $f(0,0;\,r,\xi)$, so that

$$f(r,\xi;\,s,\xi+\eta)=f(r,s;\,\eta),$$

and, due to the continuity of the function on the left, this is precisely the assertion made in the theorem.

Remark 4. In all classical and modern population (and fission) problems known, it is customary (and probably also appropriate) to assume $f(r,x;\,s,y)=0$ if either $x<0$ or $y<0$, which is an outright violation of our special assumption (5.2.23), and in the solutions of these problems the random increments $\{\xi^p\}$ are usually *not* stochastically independent, which is in sharp contrast to the solutions of the classical diffusion problems (Brownian motion), in which the stochastic independence of the 'infinitesimal' increments has been a matter of axiomatic postulation, traditionally.

Remark 5. Stationary Markoff *chains* as such,

$$f(r,x;\,s,y)=f(s-r;\,x,y),$$

even nonhomogeneous ones, were easily generalized in chapter 4 from the half-line $0<t<\infty$ to a multidimensional time variable in an octant, but the corresponding generalization of the processes offers a difficulty which we cannot readily overcome, and the difficulty is this that no order relation for the indices (5.2.1) can be then suitably introduced for which the decisive property (4) of definition 5.1.1 can be retained. Paul Lévy in defining what he terms a 'multiple Markoff process' has also encountered a certain difficulty in actually establishing the existence of his processes, and it is a difficulty of the same origin perhaps.

5.3. Length of random paths in homogeneous spaces

If we restrict the points t^p in (5.2.5) to lying in an open or closed interval $0\leqq t<a$ or $0\leqq t\leqq a$, then the resulting indices are again a directed set, and the previous constructions again apply. Also, in every case the limit space Ω: (ω) can be identified, in a familiar

manner, with the space of paths ω: $x=x(t)$ in R originating at the fixed point x^0, $0 \leqq t < a$ (or $\leqq a$) and $x(0)=x^0$.

We now take in $0 \leqq t \leqq 1$ a Markoff process which is space homogeneous and for which there is defined a non-negative Baire function

$$\rho(r, x; s, y) \geqq 0 \tag{5.3.1}$$

for $0 \leqq r < s \leqq 1$, $x, y \in R$, with

$$\rho(r, x; s, y) + \rho(s, y; t, z) \geqq \rho(r, x; t, z), \quad 0 \leqq r < s < t \tag{5.3.2}$$

(triangle property) and

$$\rho(r, Ux; s, Uy) = \rho(r, x; s, y) \tag{5.3.3}$$

(homogeneity). By theorem 5.2.2 the function

$$A(r, s; u) = \int_R e^{-u\rho(r, x; s, y)} f(r, x; s, y) dv_y, \quad u > 0 \tag{5.3.4}$$

is independent of x, and in conjunction with (5.3.2) it also implies

$$A(r, s; u)\, A(s, t; u) \leqq A(r, t; u). \tag{5.3.5}$$

Since also $0 < A(r, s; u) \leqq 1$, if we put

$$A(r, s; u) = e^{-B(r, s; u)} \tag{5.3.6}$$

then the new function satisfies

$$B(r, s; u) \geqq 0, \quad B(r, s; u) + B(s, t; u) \geqq B(r, t; u), \tag{5.3.7}$$

and thus is, for each u, a distance function on the interval $0 \leqq t \leqq 1$.

If with the indices (5.2.1) we associate the functions

$$F_\lambda(\omega) = \sum_{p=1}^{l} \rho(t^{p-1}, x(t^{p-1}); t^p, x(t^p)),$$

then these fall under theorems 5.1.2 and 5.1.3, and we are introducing the limit function $F(\omega)$ and the Laplace transforms $\phi_\lambda(u)$ and $\phi(u)$ as in these theorems. We note that the number $F(\omega)$ which is defined for almost all ω, whether it be finite or infinite, is in the nature of a length of the path, even if the underlying distance function (5.3.1) should indeed involve the variables r, s, as we have permitted it to do. But we ought to point out that we know of no case in which it does so depend without $F(\omega)$ being infinite for almost all ω.

By theorem 5.1.2 we have

$$\phi_\lambda(u) = \prod_{p=1}^{l} A(t^{p-1}, t^p; u) = \exp\left[-\sum_{p=1}^{l} B(t^{p-1}, t^p; u)\right],$$

and thus for the function $\phi(u) = E\{e^{-uF(\omega)}\}$ we can write

$$\phi(u) = e^{-C(u)}, \tag{5.3.8}$$

where $$C(u)=\sup_\lambda \sum_{p=1}^{l} B(t^{p-1}, t^p; u) \tag{5.3.9}$$

is in the nature of a total variation of $B(r, s; u)$ over the interval

$$0 \leqq r < s \leqq 1. \tag{5.3.10}$$

Also, if for each $0 < r \leqq 1$ we introduce the corresponding variation $C(r; u)$ over the interval $(0, r)$ instead of $(0, 1)$ and put $C(0; u)=0$, then $C(r; u)$ is monotonely increasing in r and we have

$$C(u) \equiv C(1; u) = \int_0^1 d_r C(r; u),$$

of course. Frequently $C(r; u)$ will be absolutely continuous in r, so that the derivative

$$D(r; u) = \frac{dC(r; u)}{dr}$$

exists a.e. in $0 \leqq r \leqq 1$, and we then have

$$\phi(u) = \exp\left[-\int_0^1 D(r; u)\, dr\right].$$

Also, if $f(r, x; s, y)$ and $\rho(r, x; s, y)$ both depend on $s-r$ only (stationarity), then we have $A(r, s; u) \equiv A(s-r; u)$, $B(r, s; u) \equiv B(s-r; u)$ and $C(s)-C(r)=C(s-r)$; and $D(r; u) \equiv D(u)$ is independent of r altogether.

Theorem 5.1.3 immediately implies the following criterion:

THEOREM 5.3.1. *If $C(u_0)=\infty$ for some $u_0 > 0$, then $C(u) \equiv \infty$, and almost all paths have infinite length. If $C(u)$ is finite, then*

$$P(S^0) = e^{-C(0+)}$$

is the measure of the set of paths of finite length, and almost all paths have finite length if and only if $C(0+)=0$.

From this we will deduce as follows:

THEOREM 5.3.2. *If for each $\epsilon > 0$ there is a $\delta > 0$ such that for $u \leqq \delta$, $s-r \leqq \delta$, we have*

$$1-A(r, s; u) \leqq \epsilon(1-A(r, s{:}\ 1)) + \epsilon(s-r), \tag{5.3.11}$$

then either almost all paths have infinite length or almost all have finite length.

Condition (5.3.11) *is fulfilled if the distance function* (5.3.1) *is bounded on R,*

$$\rho(r, x; s, y) \leqq \rho_0, \tag{5.3.12}$$

or more generally if for some $\rho_0 > 0$ *we have*

$$\int_{\rho(r,x;\,s,y)\geqq\rho_0} e^{-u\rho(r,x;\,s,y)} f(r,x;\,s,y)\,dv_y \leqq \epsilon(s-r) \tag{5.3.13}$$

for $s-r \leqq \delta \equiv \delta(\epsilon)$.

Proof. We will denote by $M_1, M_2, M_3, \ldots$, certain finite positive numbers which are independent of r, s, ϵ, δ, the latter all in $(0,1)$. By theorem 5.3.1 it suffices to prove that

$$C(v) \leqq M_1, \quad 0 < v \leqq 1, \tag{5.3.14}$$

implies

$$\lim_{u \downarrow 0} C(u) = 0. \tag{5.3.15}$$

Now (5.3.14) implies $B(r,s;\,v) \leqq M_1$, and from

$$1 - A(r,s;\,v) = 1 - e^{-B(r,s;\,v)},$$

we thus obtain

$$M_2 B(r,s;\,v) \leqq 1 - A(r,s;\,v) \leqq M_3 B(r,s;\,v). \tag{5.3.16}$$

Therefore

$$M_2 B(r,s;\,u) \leqq \epsilon(1 - A(r,s;\,1)) + \epsilon(s-r) \leqq \epsilon M_3 B(r,s;\,1) + \epsilon(s-r),$$

and hence

$$M_2 \sum_{p=1}^{l} B(t^{p-1}, t^p;\,u) \leqq \epsilon M_3 \sum_{p=1}^{l} B(t^{p-1}, t^p;\,1) + \epsilon \sum_{p=1}^{l} (t^p - t^{p-1})$$
$$\leqq \epsilon M_3 C(1) + \epsilon = \epsilon M_4.$$

Therefore, $M_2 C(u) \leqq \epsilon M_4$, which proves (5.3.15). Next we have

$$1 - A(r,s;\,u) = \int_R (1 - e^{-u\rho}) f\,dv$$
$$= \int_{\rho \leqq \rho_0} + \int_{\rho > \rho_0} = J_1(r,s;\,u) + J_2(r,s;\,u).$$

However, for $\rho \leqq \rho_0$ and $v \leqq 1$ we have

$$M_5 v\rho \leqq 1 - e^{-v\rho} \leqq M_6 v\rho, \tag{5.3.17}$$

and hence

$$1 - A(r,s;\,u) \leqq u M_7 \int_{\rho \leqq \rho_0} (1 - e^{-\rho}) f\,dv + J_2 \leqq u M_7 (1 - A(r,s;\,1)) + J_2,$$

and by assumption (5.3.13) this implies (5.3.11) as claimed.

Next, for stationary processes the reader will easily obtain the following statements from the definition of the symbols.

THEOREM 5.3.3. *If our process is also stationary then either*

$$\overline{\lim_{\epsilon \downarrow 0}} \frac{1 - A(\epsilon;\,1)}{\epsilon} = \overline{\lim_{\epsilon \downarrow 0}} \frac{B(\epsilon;\,1)}{\epsilon} = +\infty,$$

in which case $C(1)=\infty$ *and almost all paths have infinite length, or, however, we have*

$$\overline{\lim_{\epsilon\downarrow 0}}\,\frac{1-A(\epsilon;u)}{\epsilon}=\overline{\lim_{\epsilon\downarrow 0}}\,\frac{B(\epsilon;u)}{\epsilon}=D(u)<\infty,$$

the limits existing, and almost all paths have finite length.

By applying (5.3.17) and theorem 5.3.2 this easily gives the following conclusion:

THEOREM 5.3.4. *In the fully homogeneous case, if we have*

$$\rho(t;x,y)\leqq\rho_0$$

on R, *or more generally if for some* $\rho_0>0$ *we have*

$$\int_{\rho(t;\,x,\,y)\geqq\rho_0} f(t;x,y)\,dv_y=o(t),\quad t\to 0,$$

then almost all paths have infinite length or finite length depending on whether

$$\overline{\lim_{\epsilon\downarrow 0}}\,\frac{1}{\epsilon}\int_{\rho\leqq\rho_0}\rho(\epsilon;x,y)f(\epsilon;x,y)\,dv_y=\infty\quad\text{or}\quad<\infty.$$

If in E_k we are given a fully homogeneous process $\{f(t;x_j)\}$ with transform $\{e^{-t\psi(\alpha_j)}\}$, then for the ordinary distance $\rho(x;y)\equiv 2\pi|y-x|$ we have

$$A(t;u)=\int_{E_k} e^{-2\pi|x|u}f(t;x_j)\,dv_x,$$

and theorem 5.3.4 could be applied to this expression. However, by Fourier transforms this is also

$$A(t;u)=\pi^{-\frac{1}{2}(k+1)}\,\Gamma\{\tfrac{1}{2}(k+1)\}\int_{E_k}\frac{ue^{-t\psi(\alpha_j)}\,dv_\alpha}{(u^2+\alpha_1^2+\ldots+\alpha_k^2)^{\frac{1}{2}(k+1)}},$$

and an analysis of the behaviour of the difference quotient

$$1/\epsilon(1-A(\epsilon;u))$$

from the latter expression is rather easier to undertake. Under the integral sign we then have to deal with the expression

$$\frac{1-e^{-\epsilon\psi(\alpha)}}{\epsilon\psi(\alpha)}\,\psi(\alpha),$$

and since for $\epsilon\downarrow 0$ this converges majorizedly towards $\psi(\alpha)$, theorem 5.3.3 leads to the following conclusion:

THEOREM 5.3.5. *If* $\psi(\alpha_j)$ *is real valued then*

$$D(u)=\pi^{\frac{1}{2}(k+1)}\,\Gamma\{\tfrac{1}{2}(k+1)\}\int_{E_k}\frac{u\psi(\alpha_j)\,dv_\alpha}{(u^2+\alpha_1^2+\ldots+\alpha_k^2)^{\frac{1}{2}(k+1)}},$$

and almost all paths have either infinite length or finite length depending on whether

$$\int_{E_k} \frac{\psi(\alpha_1, \ldots, \alpha_k)\, dv_\alpha}{(1+\alpha_1^2+\ldots+\alpha_k^2)^{\frac{1}{2}(k+1)}} = \infty \quad \text{or} \quad < \infty.$$

In particular if $\psi(\alpha_j) = \psi(|\alpha|)$ *depends only on the distance*

$$|\alpha| = (\alpha_1^2 + \ldots + \alpha_k^2)^{\frac{1}{2}}$$

then the alternative is

$$\int_1^\infty \frac{\psi(\beta)\, d\beta}{\beta^2} = \infty \quad \text{or} \quad < \infty;$$

and thus for the stable process $\psi(\alpha_j) = |\alpha|^q$ *we have the result that irrespective of the dimension of the space we may expect infinite length if* $1 \leqq q \leqq 2$ *(from Gauss's process to Cauchy's process, inclusive), and for* $0 < q < 1$ *we may expect finite length.*

If $\psi(\alpha_j)$ *is not real valued we have at any rate the incomplete alternative that almost all paths have infinite or finite length depending on whether*

$$\int_{E_k} \frac{\operatorname{Re} \psi(\alpha_j)\, dv_\alpha}{1+|\alpha|^{k+1}} = \infty \quad \text{or} \quad \int_{E_k} \frac{|\psi(\alpha_j)|\, dv_\alpha}{1+|\alpha|^{k+1}} < \infty.$$

The following supplement might be of some interest:

THEOREM 5.3.6. *If* $\psi(\alpha_j) = \psi^G + \psi^P$ *is real valued then almost all paths have finite length if and only if the Gaussian part is missing,* $\psi^G \equiv 0$, *and also the distribution function* $F(A)$ *in the representation*

$$\psi^P(\alpha_j) = \int_{E_k'} (1 - \cos 2\pi(\alpha, \xi))\, d_\xi F(\xi) \tag{5.3.18}$$

is such that we have

$$\int_{0<|\xi|<1} |\xi|\, d_\xi F(\xi) < \infty$$

for the first power $|\xi|$, *and not only for the second power* $|\xi|^2$ *as must be the case automatically.*

Proof. In fact, for $\psi^G \not\equiv 0$ the integral is always infinite, and for (5.3.18) it has the value

$$c \int_{E_k'} (1 - e^{-2\pi|\xi|})\, d_\xi F(\xi),$$

whence the conclusion.

Finally, we will very fleetingly comment on a certain type of process which arises in the following manner. Take a continuous function $\psi(t; \alpha_j)$ for $0 \leqq t < \infty$, $\alpha \in E_k$, and assume that $e^{-u\psi(t;\, \alpha_j)}$ is a characteristic function in E_k for each t and $u > 0$. If we approximate

$$\int_r^s \psi(t; \alpha)\, dt$$

by Riemann sums it follows easily that

$$\exp\left[-\int_r^s \psi(t;\,\alpha_j)\,dt\right]$$

is again a characteristic function for $0 \leqq r < s$. Assume that it is the transform of a density (this could be generalized) and denote the latter by $f(r,s;\,x)$. Obviously $f(r,s;\cdot) * f(s,t\cdot) = f(r,t;\cdot)$, and thus $f(r,s;\,y_j - x_j)$ is a Markoff density which is space homogeneous but not stationary. It satisfies the pair of differential equations

$$\frac{\partial f(r,s;\,x)}{\partial s} = -\psi(s;\,D_1,\ldots,D_k)f,$$

$$\frac{\partial f(r,s;\,x)}{\partial r} = \psi(r;\,D_1,\ldots,D_k)f,$$

and the statement we wanted to make is as follows:

THEOREM 5.3.7. *Subject to secondary assumptions, the statements of Theorem* 5.3.5 *remain literally in force if we put now*

$$\psi(\alpha_j) \equiv \int_0^1 \psi(r;\,\alpha_j)\,dr, \quad D(u) = \int_0^1 D(r;\,u)\,dr,$$

where $D(r;\,u) = \pi^{-\frac{1}{2}(k+1)}\,\Gamma\{\tfrac{1}{2}(k+1)\}\displaystyle\int_{E_k} \frac{u\psi(r;\,\alpha_j)\,dv_\alpha}{(u^2+\alpha_3^2+\ldots+\alpha_k^2)^{\frac{1}{2}(k+1)}}.$

5.4. Euclidean stochastic processes and their characteristic functionals

We take a directed family of *Euclidean* spaces $\{\Omega_\lambda\}$ with connecting transformations $f_{\lambda\mu}$ as in section 5·1, giving rise to a projective limit $\Omega\colon(\omega)$, and we assume that each $f_{\lambda\mu}$ is of the form

$$x_p = \sum_{q=1}^{m} c_{pq} y_q, \quad p = 1,\ldots,l, \tag{5.4.1}$$

where $\Omega_\lambda \equiv E_l\colon(x_p)$ and $\Omega_\mu \equiv E_m\colon(y_q)$. We note that

$$\text{rank of matrix } \{c_{pq}\} = l, \tag{5.4.2}$$

since $f_{\lambda\mu}$ is a mapping into all of Ω_λ, as always assumed.

With each given

$$\Omega_\lambda\colon(\omega_\lambda) \equiv E_l\colon(x_p), \tag{5.4.3}$$

we now associate its Fourier analytic dual

$$\hat{\Omega}_\lambda\colon(\hat{\omega}_\lambda) \equiv E_l\colon(\alpha_p), \tag{5.4.4}$$

and correspondingly with each 'inverse' mapping (5.4.1) the contragradient 'forward' mapping

$$\beta_q = \sum_{p=1}^{l} c_{pq}\alpha_p, \quad q=1,\ldots,m, \tag{5.4.5}$$

for which we also write $$\hat{\omega}_\mu = g_{\mu\lambda}\hat{\omega}_\lambda, \tag{5.4.6}$$

and we note the following properties, the last of which is a consequence of (5.4.2):

LEMMA 5.4.1. *We have*

$$g_{\nu\mu}g_{\mu\lambda} = g_{\nu\lambda} \quad \text{for} \quad \nu>\mu>\lambda, \tag{5.4.7}$$

and $$\sum_{p=1}^{l} x_p\alpha_p = \sum_{q=1}^{m} y_q\beta_q, \tag{5.4.8}$$

so that $$(f(y),\alpha) = (y, g(\alpha)), \tag{5.4.9}$$

and if for two points $\hat{\omega}_\lambda^1$, $\hat{\omega}_\lambda^2$ *and* $\mu>\lambda$, *we have* $g_{\mu\lambda}(\hat{\omega}_\lambda^1) = g_{\mu\lambda}(\hat{\omega}_\lambda^2)$ *then* $\hat{\omega}_\lambda^1 = \hat{\omega}_\lambda^2$.

DEFINITION 5.4.1. The (*'forward'*) *projective limit*

$$\hat{\Omega}\colon (\hat{\omega}) = \lim_\lambda \{\hat{\Omega}_\lambda\colon (\hat{\omega}_\lambda)\}, \quad \lambda \in \Lambda, \tag{5.4.10}$$

is defined as follows. A point $\hat{\omega}$ of it is any *maximal* collection of points

$$\{\hat{\omega}_\lambda\} \tag{5.4.11}$$

(called its '*components*') and a subset Λ^0 of Λ which depends on $\hat{\omega}$ such that (i) for each $\lambda \in \Lambda^0$ the collection contains one and only one point $\hat{\omega}_\lambda$ of $\hat{\Omega}_\lambda$, (ii) if $\lambda_1, \lambda_2 \in \Lambda^0$ (where $\lambda_1 = \lambda_2$ is permissible) and $\lambda_3 > \lambda_1, \lambda_2$ then $\lambda_3 \in \Lambda^0$ and (iii) for the corresponding components we have

$$g_{\lambda_3\lambda_1}(\hat{\omega}_{\lambda_1}) = g_{\lambda_3\lambda_2}(\hat{\omega}_{\lambda_2}) \equiv \hat{\omega}_{\lambda_3}. \tag{5.4.12}$$

THEOREM 5.4.1. (i) *If two points of* $\hat{\Omega}$ *have one component in common then they are identical, and every point* $\hat{\omega}_{\lambda_0}$ *of every* $\hat{\Omega}_{\lambda_0}$ *is the component of a point* $\hat{\omega}$ *which can be described as follows. For every* $\lambda > \lambda_0$ *the point* $\hat{\omega}_\lambda = g_{\lambda\lambda_0}(\hat{\omega}_{\lambda_0})$ *is a component of it and if for an index* $\mu < \lambda$, $\mu < \lambda_0$ *and a point* $\hat{\omega}_\mu$ *it so happens that* $g_{\lambda\mu}(\hat{\omega}_\mu) = g_{\lambda\lambda_0}(\hat{\omega}_{\lambda_0})$ *then it is likewise a component.*

(ii) *The mapping* $\hat{\omega} \equiv g_{\infty\lambda_0}(\hat{\omega}_{\lambda_0})$ *is a transformation of* Ω_{λ_0} *into* $\hat{\Omega}$ *such that*

$$g_{\infty\lambda} = g_{\infty\mu}g_{\mu\lambda} \tag{5.4.13}$$

for $\lambda < \mu$.

(iii) *For any* $n+1$ *points* $\hat{\omega}^r$ *in* $\hat{\Omega}$, $r=0,1,2,\ldots,n$; $n \geqq 1$, *there is an index* λ *for which the components* $\hat{\omega}_\lambda^r$ *exist simultaneously, and* $\hat{\Omega}$ *can be made into a real vector space in such a manner that for any such combination of points and an index* λ *the relation*

$$\hat{\omega}^0 = a_1 \hat{\omega}^1 + \ldots + a_n \hat{\omega}^n \tag{5.4.14}$$

in $\hat{\Omega}$ *implies the relation*

$$\hat{\omega}^0_\lambda = a_1 \hat{\omega}^1_\lambda + \ldots + a_n \hat{\omega}^n_\lambda$$

in $\hat{\Omega}_\lambda$.

This can be proved readily with the aid of lemma 5.4.1. Also, the original projective limit Ω can be likewise trivially made into a vector space in such a manner that the given spaces Ω_λ are vector projections of it. If now we introduce the notation

$$(\omega_\lambda, \hat{\omega}_\lambda) = \sum_{p=1}^{l} x_p \alpha_p \tag{5.4.15}$$

and recall the consistency relations (5.4.8), then the following statement is easily proven:

THEOREM 5.4.2. *On the product vector space* $\Omega \times \hat{\Omega}$ *there exists a bidistributive functional* $(\omega, \hat{\omega})$ *such that we have*

$$(\omega, \hat{\omega}) = (\omega_\lambda, \hat{\omega}_\lambda) \tag{5.4.16}$$

for every λ *for which the component* $\hat{\omega}_\lambda$ *exists.*

If we take a nonempty *directed* subset Λ' of Λ and introduce the corresponding limit spaces

$$\Omega' = \lim_\lambda \Omega', \quad \hat{\Omega}' = \lim_\lambda \Omega'_\lambda; \quad \lambda \in \Lambda', \tag{5.4.17}$$

then $\hat{\Omega}'$ can be mapped vector-isomorphically into a part of $\hat{\Omega}$ by assigning to a point $\hat{\omega}'$ the point $\hat{\omega}$ having with it any one component in common for $\lambda \in \Lambda'$. The corresponding statement for the inverse limits is also true, but nontrivial, and if in particular we choose for Λ' a directed sequence $\{\lambda_n\}$ then this asserts the sequential maximality needed in theorem 5.1.1.

THEOREM 5.4.3. *With each* $\omega' \in \Omega'$ *there can be associated at least one point* $\omega \in \Omega$ *having the same components as* ω' *for* $\lambda \in \Lambda'$.

Proof. For fixed ω' and variable $\hat{\omega}'$ we view $(\omega', \hat{\omega}')$ as a distributive functional on $\hat{\Omega}'$. Now $\hat{\Omega}'$ is a vector subspace of $\hat{\Omega}$, and since we are not concerned with 'completeness' or 'closure' of the spaces and functionals at all, there is a very simple general theorem available (based on the axiom of choice) stating that any distributive operation from a vector space $\hat{\Omega}'$ to a vector space of values V (which for us are real numbers) can be extended to any vector space $\hat{\Omega} \supset \hat{\Omega}'$. We pick one such extension and denote it by $(\omega', \hat{\omega})$. Now, if ν is any index in the large set Λ, then for the points $\hat{\omega}$ having components $\hat{\omega}_\nu = (\gamma_1, \ldots, \gamma_n)$ in $\hat{\Omega}_\nu \equiv E_n$: (γ) the functional must have the familiar form

$$z_1\gamma_1 + \ldots + z_n\gamma_n,$$

and if now we put $\omega_\nu = (z_1, \ldots, z_n)$ then $\{\omega_\nu\} \equiv \omega$ is a point having the property asserted.

DEFINITION 5.4.2. We call a stochastic family Euclidean if each given probability space has a Euclidean structure

$$\Omega_\nu \equiv E_l; \quad \mathscr{S}_\lambda \equiv \mathscr{A}_l; \quad P_\lambda(S) = F_l(A), \tag{5.4.18}$$

and if all connecting transformations $f_{\lambda\mu}$ have the *affine* structure (5.4.1), with origin going into origin.

By theorem 5.4.3 we may now assert as follows:

THEOREM 5.4.4. *Any Euclidean stochastic family has the Kolmogoroff property so that a limiting probability space*

$$\{\Omega; \mathscr{S}; P\} \tag{5.4.19}$$

exists, and the resulting stochastic process will be called 'Euclidean'.

For each λ, we now introduce the characteristic function

$$\phi_\lambda(\alpha_j) = \int_{E_l} e^{-2\pi i(x,\, \alpha)}\, dF_l(x), \tag{5.4.20}$$

and, in keeping with (5.4.15), we write for this also

$$\phi_\lambda(\hat{\omega}_\lambda) = \int_{\Omega_\lambda} e^{-2\pi i(\omega_\lambda,\, \hat{\omega}_\lambda)}\, dP_\lambda. \tag{5.4.21}$$

But now the following statements become obvious:

THEOREM 5.4.5. (i) *Our Euclidean process can also be characterized 'dually' by a directed family of data*

$$\hat{\Omega}_\lambda \equiv E_l: (\alpha_p); \quad \phi_\lambda(\hat{\omega}_\lambda) = \phi_\lambda(\alpha_1, \ldots, \alpha_l), \tag{5.4.22}$$

with connecting affine transformations (5.4.6) *of the form* (5.4.5) *for which* (5.4.2) *and* (5.4.7) *hold; the functions* $\phi_\lambda(\alpha_j)$ *being characteristic functions which are linked by the identities*

$$\phi_\lambda(\alpha_p) = \phi_\mu(\beta_q) \equiv \phi_\mu(g_{\mu\lambda}(\alpha)), \tag{5.4.23}$$

or written differently $\quad \phi_\lambda(\hat{\omega}_\lambda) = \phi_\mu(g_{\mu\lambda}(\hat{\omega}_\lambda)),$

for $\lambda < \mu$.

(ii) *On the limit space* $\hat{\Omega}$ *there is given a function* $\phi(\hat{\omega})$ *such that for any component* $\hat{\omega}_\lambda$ *of any* $\hat{\omega}$ *we have*

$$\phi(\hat{\omega}) = \phi_\lambda(\hat{\omega}_\lambda), \tag{5.4.24}$$

and we can also write $$\phi(\hat{\omega}) = \int_{\Omega} e^{-2\pi i(\omega, \hat{\omega})} dP(\omega) \equiv \int_{\Omega_\lambda} e^{-2\pi i(\omega_\lambda, \hat{\omega}_\lambda)} dP_\lambda(\omega), \tag{5.4.25}$$

the function $(\omega, \hat{\omega})$, *and hence also* $e^{-2\pi i(\omega, \hat{\omega})}$ *being a Baire function on* (5.4.19) *for each* $\hat{\omega}$.

DEFINITION 5.4.3. This function $\phi(\hat{\omega})$ will be called the *characteristic functional* of the given Euclidean stochastic process.

5.5. Random functions

DEFINITION 5.5.1. On a general point set T: (t), a family of its *subsets* $D = \{\Delta\}$ will be called *admissible* if corresponding to any finite numbers of its elements

$$\Delta_1, \Delta_2, \ldots, \Delta_l \tag{5.5.1}$$

(which we will call a *union*) there is another finite number

$$\Delta_1', \Delta_2', \ldots, \Delta_m' \tag{5.5.2}$$

in which any two Δ_r', Δ_s' are disjoint for $r \neq s$ (which we then call a *sum*) such that each Δ_p is a point set sum of certain Δ_q' which we will indicate by writing

$$\Delta_p = \sum_{q=1}^{q_p} \Delta_{pq}', \quad p = 1, \ldots, l. \tag{5.5.3}$$

We now take a function $X(\Delta)$ on D with real values say, and we call it additive, if every relation (5.5.3) implies

$$X(\Delta_p) = \sum_{q=1}^{q_p} X(\Delta_{pq}'). \tag{5.5.6}$$

An admissible family D arises if we take for Δ any and every single point t of T (only), and in this case an 'additive' function $X(\Delta)$ is any point function $X(t)$ whatsoever. On the other hand, if D is a σ-algebra of sets then our $X(\Delta)$ is an additive 'set' function in the familiar sense; and finally we may take, for instance, on $T = \{0 < t < \infty\}$, for D the family of intervals $\Delta_{\alpha\beta} = (\alpha < t \leqq \beta)$, either for all real end-points α, β, or only rational ones, or even only diadic ones $m/2^n$, with $X(\Delta)$ being then an interval function in the usual sense.

Remark 1 (*Important*). A function $X(\Delta)$ on 'Δ', whether real-valued or later random-valued or Banach-valued, will *always* be intended to be additive as in definition 5.5.1, and sometimes, but not always, we will recall this for emphasis.

Given D, we introduce a directed set $\Lambda = \{\lambda\}$ in the following manner. An index is any sum

$$\lambda = (\Delta_1, \Delta_2, \ldots, \Delta_l), \tag{5.5.7}$$

and if we are given another index

$$\mu = (\Delta_1', \Delta_2', \ldots, \Delta_m'), \tag{5.5.8}$$

then we put $\lambda < \mu$ if each Δ_p of λ is a sum (5.5.3) of elements of μ, with some elements Δ_r' left over perhaps. Thus, if we view λ as an (incomplete) partitioning of T then $\mu > \lambda$ means that μ refines λ and adds some elements, perhaps. This is indeed a directed set, and, in fact, the crucial property (4) of definition 5.1.1 is fulfilled because for any two indices $\lambda_1 = (\Delta_{p_1}^1)$, $\lambda_2 = (\Delta_{p_2}^2)$ we can form the union $\{\Delta_{p_1}^1, \Delta_{p_2}^2\}$, and by definition 5.5.1 there exists then a sum $\mu = (\Delta_q')$ such that $\lambda_1 < \mu$, $\lambda_2 < \mu$ simultaneously. Frequently the set Λ will contain a subset Λ' which will be a directed set likewise, and thus be 'admissible' in the forthcoming theorem itself. For instance, if the set T itself is a sum (5.5.1), then the subset Λ' of indices which 'refine' (5.5.1) is of this kind, and a very suitable directed set of indices it is.

If for a function $X(\Delta)$ we associate with two indices (5.5.7), (5.5.8) the values

$$\left.\begin{aligned} x_1 = X(\Delta_1), \ldots, x_l = X(\Delta_l), \\ y_1 = X(\Delta_1'), \ldots, y_m = X(\Delta_m'), \end{aligned}\right\} \tag{5.5.9}$$

then corresponding to (5.5.6) we have

$$x_p = y_{p1} + \ldots + y_{pq_p}, \quad p = 1, \ldots, l, \tag{5.5.10}$$

and if we put this in the form

$$f_{\lambda\mu}\colon x_p = \sum_{q=1}^{m} c_{pq} y_q, \quad p = 1, \ldots, l, \tag{5.5.11}$$

where c_{pq} are certain constants $= 1$ or 0, then these transformations satisfy requirements (5.1.3) and (5.1.4), among others. We can now apply the results of section 5.4, and we first state the following theorem, the first half of which is the easy one:

THEOREM 5.5.1. *If on an admissible family D we are given an additive function $\tilde{X}(\Delta)$,*

$$\tilde{X}(\Delta_1 + \ldots + \Delta_l) = \tilde{X}(\Delta_1) + \ldots + \tilde{X}(\Delta_l), \tag{5.5.12}$$

whose values $\tilde{X}$ are random variables on some given probability space

$$\{\Omega; \mathscr{S}; P\}, \tag{5.5.13}$$

and if with every index in Δ *(or only some* Λ'*) we associate the Euclidean space*

$$\{\Omega_\lambda \equiv E_l;\ \mathscr{S}_\lambda \equiv \mathscr{A}_k;\ P_\lambda \equiv F_l\}, \tag{5.5.14}$$

where F_l *is the joint distribution function of the* l *random variables* $x_1 = \tilde{X}(\Delta_1), \ldots, x_l = \tilde{X}(\Delta_l)$; *then these Euclidean spaces constitute a stochastic family for the transformations* (5.5.11).

Conversely, if we are given D *and for every* λ *a probability measure* $F_l(A)$ *in* E_l, *and if they are linked by the given transformations* (5.5.11) *on* D, *then* (5.5.14) *constitutes a stochastic process whose limit space* Ω *consists of all numerical additive functions* $\omega = X(\Delta)$,

$$X(\Delta_1 + \ldots + \Delta_l) = X(\Delta_1) + \ldots + X(\Delta_l), \tag{5.5.15}$$

and for each λ *the set function* $F_l(A)$ *is the probability for the* l *values* $x_1 = X(\Delta_1), \ldots, x_l = X(\Delta_l)$ *to lie in the Borel set* A.

Remark 2. In fulfilling requirement (5.5.12) we may admit exceptional sets of measure 0, the sets varying with $(\Delta_1, \ldots, \Delta_l)$; for the constructed random function $X(\Delta)$, however, (5.5.15) is fulfilled without exception. Thus any random function may be replaced by another having the same joint distribution functions for which (5.5.15) is fulfilled without exceptional sets, but the new random variables are on a probability space altogether different from the original one.

Also in what follows the symbol $X(\Delta)$ without the symbol '$\sim$' will frequently denote an arbitrary (additive) random function on D and not necessarily the one constructed in theorem 5.5.1.

Remark 3. If the value of the function $X(\Delta)$ is a k-dimensional vector, then not much need be changed, provided we replace the dimension l in (5.5.14) by kl and then take for $F_{kl}(A)$ the joint distribution function of the various components of the l vectors $X(\Delta_p)$; and for $k=2$ the case of complex-valued random functions is fully included in this way.

Definition 5.5.2. For given D a step function on an index λ is a real function on T which for $t \in \Delta_p$ has a constant value h_p, $p = 1, \ldots, l$, and for $t \in T - (\Delta_1 + \ldots + \Delta_l)$ has the value 0, and thus can also be written as

$$h(t) = \sum_{p=1}^{l} h_p \omega_{\Delta_p}(t), \tag{5.5.16}$$

where $\omega_\Delta(t)$ is the characteristic function of the set Δ.

For given λ, all step functions on it are a vector space. Also for $\mu > \lambda$ a step function on λ is also one on μ, and since $\{\lambda\}$ is directed we obtain the following statement:

Lemma 5.5.1. *All step functions together constitute a vector space which we will denote by* L_0.

If $X(\Delta)$ is a random function on D, then for a given step function (5.5.16) the sum

$$\sum_{p=1}^{l} h_p X(\Delta_p) \tag{5.5.17}$$

is independent of the special index λ used, and depends therefore on the step function $h(t)$ as a point function on T only.

Definition 5.5.3. For any $h \in L_0$, we denote this value by

$$\int_T h(t)\, dX(t) \equiv \int_T h(t)\, dX \tag{5.5.18}$$

and view it as a *Stieltjes integral* then.

If now we put $h_p = \alpha_p$, $F(\Delta_p) = x_p$, then (5.5.17) is the sum

$$x_1\alpha_1 + \ldots + x_l\alpha_l,$$

and theorem 5.4.5 implies the following one:

Theorem 5.5.2. *The space $\hat{\Omega}$: $(\hat{\omega})$ dual to the space Ω of theorem 5.5.1 can be aptly represented by the vector space L_0, so that*

$$(\omega, \hat{\omega}) = \int_T h(t)\, dX \tag{5.5.19}$$

and

$$\phi(\hat{\omega}) \equiv \phi(h) = E\left\{\exp\left[-2\pi i \int_T h(t)\, dX\right]\right\}, \tag{5.5.20}$$

and this has the same value as

$$E\left\{\exp\left[-2\pi i \int_T h(t)\, d\tilde{X}\right]\right\},$$

where $\tilde{X}(\Delta)$ is any other random-valued additive functions on D having the same joint distribution functions $F_l(A)$ as $X(\Delta)$ itself.

Remark 4. If $X(\Delta)$ is vector-valued, then nothing else is changed if we also make $h(t)$ vector-valued and in (5.5.17) interpret $h \cdot X$ as $h'X' + h''X'' + \ldots + h^{(k)}X^{(k)}$, and for $k = 2$ we may also write for this $h\bar{X} + \bar{h}X$, where $h = \frac{1}{2}(h' + ih'')$, $X = \frac{1}{2}(X' + iX'')$, so that then

$$(\omega, \hat{\omega}) = \int_T h(t)\, d\bar{X} + \int_T \overline{h(t)}\, dX.$$

We now assume that there is given on D a finitely additive measure $\nu(\Delta) \geqq 0$,

$$\nu(\Delta_1 + \ldots + \Delta_l) = \nu(\Delta_1) + \ldots + \nu(\Delta_l),$$

and we are introducing the following definition:

Definition 5.5.4. Given $\{D, \nu(\Delta)\}$, we call the random function $X(\Delta)$ a *homogeneous process*, if there is a function $\psi(\alpha)$, $-\infty < \alpha < \infty$, as in a subdivisible process, such that the characteristic function of the distribution $F_l(A)$ in (5.5.14) is

$$\phi_l(\alpha_1, \ldots, \alpha_l) = \exp[-\nu(\Delta_1)\psi(\alpha_1) - \ldots - \nu(\Delta_l)\psi(\alpha_l)]. \quad (5.5.21)$$

The random variables $X(\Delta_1), \ldots, X(\Delta_l)$ are then obviously stochastically independent for any disjoint $\Delta_1, \ldots, \Delta_l$ in D. Also, if $T = (0 < t < \infty)$ and $\Delta \equiv \Delta_{\alpha\beta} = (\alpha < t \leqq \beta)$, then this homogeneity is the same as the (full) homogeneity of a Markoff process if we interpret $X(\Delta)$ as the increments $\xi = x(\beta) - x(\alpha)$ of a path $x(t)$, as was already stated in remark 1 of section 5.2.

For a step function (5.5.16) the sum

$$\sum_{p=1}^{l} \psi(h_p)\, v(\Delta_p)$$

can be designated as the symbol

$$\int_T \psi(h(t))\, dv, \quad (5.5.22)$$

and we now have the following statement:

THEOREM 5.5.3. *For a homogeneous process $X(\Delta)$ the characteristic functional has the value*

$$\psi(h) = \exp\left[-\int_T \psi(h(t))\, dv_t\right], \quad h \in L_0, \quad (5.5.23)$$

so that for instance for $\psi(\alpha) = \alpha^2$ [symmetric Gaussian homogeneous process] we have

$$\phi(h) = \exp\left[-\int_T h(t)^2\, dv\right], \quad (5.5.24)$$

and for $\psi(\alpha) = |\alpha|^\rho$, $0 < \rho < 2$ [symmetric Lévy homogeneous process] we have

$$\phi(h) = \exp\left[-\int_T |h(t)|^\rho\, dv\right]. \quad (5.5.25)$$

Proof. By (5.5.20) and stochastic independence we have

$$\phi(h) = E\{e^{-2\pi i \Sigma h_p X(\Delta_p)}\} = \prod_{p=1}^{l} E\{e^{-2\pi i h_p X(\Delta_p)}\}$$
$$= \prod_{p=1}^{l} e^{-v(\Delta_p)\psi(h_p)} = \exp\left[-\sum_{p=1}^{l} \psi(h_p)\, v(\Delta_p)\right],$$

and this is the integral (5.5.23), as claimed.

Remark 5. If $X(\Delta)$ is a k-vector, and correspondingly

$$\psi(\alpha) \equiv \psi(\alpha^1, \ldots, \alpha^k),$$

then (5.5.23) generalizes to

$$\phi(h^1, \ldots, h^k) \equiv \exp\left[-\int_T \psi(h^1(t), \ldots, h^k(t))\, dv\right], \tag{5.5.26}$$

and if, for instance,

$$\psi(\alpha^1, \ldots, \alpha^k) = \sum_{p,\, q=1}^{k} c_{pq}\alpha^p \alpha^q \geqq 0, \tag{5.5.27}$$

then $$\phi(h^1, \ldots, h^k) \equiv \exp\left[-\int_T \Sigma c_{pq} h^p(t)\, h^q(t)\, dv(t)\right]. \tag{5.5.28}$$

Remark 6. If all Δ's are individual points so that our function is a proper point function $X(t)$, then $\lambda = (t_1, \ldots, t_l)$, and (5.5.19) is a finite discrete sum $\sum_{p=1}^{l} h_p X(t_p)$ according to our general definition 5.4.3. If now T is the continuous interval $0 < t \leqq 1$ say, then one would prefer having an integral

$$\int_0^1 h(t)\, X(t)\, dt \tag{5.5.29}$$

instead, and we would like to point out that this can be so obtained by adjusting the stochastic process $X(t)$ itself rather than modifying the definition 5.4.3 as such. In fact, if the integral (5.5.29) exists for such simple noncontinuous functions as are step functions on intervals $\Delta_{\alpha\beta}$, then by putting $h(t) = \omega_{\Delta_{\alpha\beta}}(t)$ we arrive at random variables

$$\int_\alpha^\beta X(t)\, dt \equiv Y(\Delta_{\alpha\beta}).$$

They constitute a random interval function, and for a step function $h(t)$ the integral (5.5.29) can now be written as the Stieltjes integral

$$\int_0^1 h(t)\, dY$$

according to our general definition. If, therefore, $X(t)$ is such that 'averaged' values $Y(\Delta)$ are a fair substitute for it (more general random functions $X(t)$ would hardly be sufficiently 'observable' at individual points t to be of interest stochastically) then the expression

$$E\left\{\exp\left[-2\pi i \int_0^1 h(t)\, X(t)\, dt\right]\right\}$$

may be claimed to be the characteristic functional for $X(t)$ itself, indirectly.

5.6. Generating functionals

If a joint distribution function $F_l(A)$ of random variables $X_1, \ldots, X_l$ is 0 outside the closure of a proper cone C_l, then for $(z_1, \ldots, z_l)$ in the (closed dual) cone Z_l (or more generally for complex (z_j) whose real parts lie in Z_l) we can introduce the completely monotone function

$$\Phi(z_1, \ldots, z_k) = \int_{\bar{C}_l} e^{-(z_1 x_1 + \cdots + z_k x_k)} \, dF_l(x),$$

called the *generating function* of F_l, which by its real values already determines F_l, and which for the complex values $z_j = 2\pi i \alpha_j$ is the characteristic function itself. In particular, if $X_1, \ldots, X_l$ are all $\geqq 0$ (as they are deemed to be in 'population' problems), then C_l is the standard octant $x_1 > 0, \ldots, x_l > 0$, but we do not demand that our octants be so normalized.

If in the case of a Euclidean stochastic process each F_l is 0 outside a closed cone $\bar{C}_l$, and if for $\lambda < \mu$ we have

$$f_{\lambda\mu}^{-1}(\bar{C}_l) = \bar{C}_m,$$

then we also have $$\Phi_\mu(g_{\mu\lambda}(z)) = \phi_\lambda(z).$$

For the points $\hat{\omega}$ in a certain 'cone' of the space $\hat{\Omega}$ we can then introduce a certain function $\Phi(\hat{\omega})$ having the form

$$\Phi(\hat{\omega}_\lambda) = \int_{\Omega_\lambda} e^{-2\pi(\omega_\lambda, \hat{\omega}_\lambda)} \, dP_\lambda(\omega_\lambda),$$

and we term it a *generating functional*. For a random function $X(\Delta)$ it has the form

$$\Phi(h) = E\left\{\exp\left(-2\pi \int_T h(t)\, dX\right)\right\},$$

naturally.

Theorem 4.2.3 implies the following one:

THEOREM 5.6.1. *If $\chi(\xi_1, \ldots, \xi_k)$ is a completely monotone function in a proper cone C_k, and*

$$\xi_j = \psi_j(\eta_1, \ldots, \eta_l), \quad j = 1, \ldots, k,$$

is a completely monotone mapping from another C_l into C_k, then

$$\frac{\chi(c_1 + \psi_1(\eta), \ldots, c_k + \psi_k(\eta))}{\chi(c_1, c_2, \ldots, c_k)}$$

is a generating function for any $(c_1, \ldots, c_k)$ in C_k, and if $\chi(\xi_j)$ is bounded in C_k, then (c_j) may also lie on the boundary of C_k.

Another, perhaps more important criterion based on it, is as follows.

THEOREM 5.6.2. *Given the data of theorem* 5.6.1, *if* $\zeta=\lambda(\eta_1, \ldots, \eta_l)$ *is a further completely monotone mapping from* C_l *to* $0<\zeta<\infty$, *and* $(\gamma_1, \ldots, \gamma_k)$ *is also in* C_k, *then the function*

$$\frac{1}{\lambda(\eta)}[\chi(c_1+\psi_1, \ldots, c_k+\psi_k)-\chi(c_1+\psi_1+\gamma_1\lambda, \ldots, c_k+\psi_k+\gamma_k\lambda)] \tag{5.6.1}$$

is completely monotone and bounded in C_l, *and thus a generating function except for a numerical factor.*

Proof. If we introduce the representation

$$\chi(\xi_1, \ldots, \xi_k)=\int_{U_k} e^{-(u_1\xi_1+\cdots+u_k\xi_k)}\,d\rho(u),$$

then the function (5.6.1) is

$$\int_{U_k} e^{-(u,\,c+\psi)}\left[\frac{1-e^{-(u,\gamma)\lambda(\eta)}}{\lambda(\eta)}\right]d\rho(u), \tag{5.6.2}$$

and (u,γ) is $\geqq 0$. If $(u,\gamma)=0$ then the bracket is $=1$, and thus completely monotone, and if $(u,\gamma)>0$, then the bracket is

$$(u,\gamma)\,.\int_0^1 e^{-(u,\,\gamma)\,t\lambda(\eta)}\,dt,$$

and thus completely monotone. Also, $e^{-(u,\,c+\psi)}\equiv e^{-(u,\,c)}\,e^{-(u,\,\psi)}$ is com pletely monotone for u in U_k, and thus the entire function (5.6.2) is so. Finally, for $\eta_1=\ldots=\eta_l=0$, (5.6.2) has at most the value

$$\int_{U_k} e^{-(u_1c_1+\cdots+u_kc_k)}\,(u_1\gamma_1+\ldots+u_k\gamma_k)\,d\rho(u),$$

which is

$$-\Sigma_j\gamma_j\left.\frac{\partial\chi}{\partial\xi_j}\right|_{\xi=c},$$

and thus finite for $c\in C_k$.

By the use of theorem 5.6.2 it can be shown that for any completely monotone function $\chi(t)$ in $0\leqq t<\infty$ with $\chi(0+)=1$ and $\chi(+\infty)=0$ the expression

$$\Phi(h)\equiv\int_{0+}^{\infty}\frac{-(1-\theta(t))}{1-\theta(t)\,e^{-h(t)}}\,d_t\chi\left(\int_{0+}^{t}(1-\theta(t)\,e^{-h(t)})\,a(t)\,dt\right)$$

for $a(t)\geqq 0$ and $0\leqq\theta(t)\leqq\theta_0<1$ is the generating functional of an interval function $X(\Delta_{\alpha\beta})$ on the intervals $(0\leqq)\alpha<t\leqq\beta(<\infty)$, and for $\theta(t)=\text{const.}$ it was introduced (in its finite dimensional versions) by J. Neyman in a study on accident proneness.

CHAPTER 6

ANALYSIS OF STOCHASTIC PROCESSES

6.1. Basic operations with characteristic functionals

To operate with characteristic functionals on a given dual space $\hat{\Omega}$ means to generate a functional $\phi(\hat{\omega})$ from given data, especially from other functionals, and by theorem 5.4.5 this amounts to constructing a family of characteristic functions $\{\phi_\lambda(\alpha_p)\}$ for which the consistency relations

$$\phi_\lambda(\alpha_p) = \phi_\mu(g_{\mu\lambda}(\alpha)) = \phi_\mu(\beta_q) \tag{6.1.1}$$

have to be verified, but also nothing else. For instance, it follows immediately that a product $\prod_{n=1}^{N} \phi^{(n)}(\hat{\omega})$ of characteristic functionals, or a sum

$$\sum_{n=1}^{N} \gamma_n \phi^{(n)}(\hat{\omega}), \quad \gamma_n \geqq 0, \quad \sum_{n=1}^{N} \gamma_n = 1,$$

or $\lim_{n\to\infty} \phi^{(n)}(\hat{\omega})$, is again a functional, and if, for instance, on the same $\{D, \Lambda'\}$ we are given two functions $X^1(\Delta)$, $X^2(\Delta)$ with random values in the same probability space, and if the two sets of random variables $\{X^1(\Delta)\}$, $\{X^2(\Delta)\}$ are stochastically independent (definition 3.7.6), then for the sum function $X^1(\Delta) + X^2(\Delta)$ the characteristic functional is the product of the two given ones.

Definition 6.1.1. We say that a Euclidean stochastic process is *symmetric, or Gaussian*, or *Poisson*, etc., if each $\phi_\lambda(\alpha_p)$ is so. Thus, for instance, a Gaussian process is one for which

$$\phi(\hat{\omega}) \equiv \exp(iQ_1(\hat{\omega}) - Q_2(\hat{\omega})), \tag{6.1.2}$$

where $Q_1(\hat{\omega})$ is a real distributive functional on $\hat{\Omega}$ and $Q_2(\hat{\omega})$ is a real non-negative symmetric bi-distributive functional on $\hat{\Omega} \times \hat{\Omega}$; and if $Q_1(\hat{\omega}) = 0$, then the process is symmetric Gaussian.

We say that the process is *Gauss-Poisson* if for each λ we have $Q_\lambda(\alpha_p) = e^{-\psi_\lambda(\alpha_p)}$, where

$$\psi_\lambda(\alpha) = \psi_\lambda^B + \psi_\lambda^G + \psi_\lambda^P, \tag{6.1.3}$$

as in (3.4.13).

Note that if a function $X(\Delta)$ is a homogeneous process as in definition 5.5.4, then it is a particular case of a Gauss–Poisson process as just defined, but only a rather particular case of it.

Relations (6.1.1) are equivalent with

$$\psi_\lambda(\alpha) = \psi_\mu(\beta), \tag{6.1.4}$$

and if we introduce new functions

$$e^{-\tilde{\psi}_\lambda(\alpha)} = \int_0^\infty e^{-t\psi_\lambda(\alpha)}\, d\rho(t),$$

then (6.1.4) implies $\tilde{\psi}_\lambda(\alpha) = \tilde{\psi}_\mu(\beta)$, and hence the following conclusion:

THEOREM 6.1.1. *The general rules of subordination (theorem* 4.3.1 *and others) also apply to characteristic functionals, so that in particular no symmetric Gaussian process on* $\hat{\Omega}$ *is subordinate to any Gauss–Poisson process.*

DEFINITION 6.1.2. For any admissible family D, a complex function $K(\Delta, \Delta')$ on $D \times D$ is a positive-definite kernel if for any complex step function (5.5.16) we have

$$\left.\begin{aligned} \sum_{p,q=1}^{l} h_p \bar{h}_q K(\Delta_p, \Delta_q) &\equiv \int_T \int_T h(t)\, \overline{h(\tau)}\, d^2K(t,\tau) \geqq 0, \\ K(\Delta', \Delta) &= \overline{K(\Delta, \Delta')}. \end{aligned}\right\} \tag{6.1.5}$$

We call it a *Hilbert kernel* if it is an inner product

$$K(\Delta, \Delta') \equiv (X(\Delta), X(\Delta'))$$

for an (additive) function $X(\Delta)$ with values in some (perhaps non-separable) Hilbert space, and we call it a *covariance kernel* if the Hilbert space can be chosen as an L_2-space over a probability measure space, that is,

$$K(\Delta, \Delta') \equiv E\{X(\Delta)\, \overline{X(\Delta')}\} \equiv \int_\Omega X(\Delta, \omega)\, \overline{X(\Delta', \omega)}\, dP(\omega).$$

Let $X(\Delta)$ be a Gaussian process with characteristic functional $\exp(-Q_2(\hat{\omega}))$. If $\hat{\omega}$ is a step function (5.5.16), then, except for the factor $4\pi^2$, $Q_2(\hat{\omega})$ is

$$\begin{aligned} E\left\{\left(\int_T h(t)\, dX(t)\right)^2\right\} &\equiv E\left\{\left(\sum_{p=1}^{l} h_p X(\Delta_p)\right)^2\right\} \\ &\equiv \sum_{p,q=1}^{l} h_p h_q K(\Delta_p, \Delta_q), \end{aligned} \tag{6.1.6}$$

where $K(\Delta, \Delta') = E\{X(\Delta)\, X(\Delta')\}$. Also in the case of a vector function $X_j(\Delta)$, $j = 1, \ldots, k$, (6.1.6) is to be replaced by

$$\sum_{p,q,j,j'} h_{pj} h_{qj'} K_{jj'}(\Delta, \Delta'),$$

where $$K_{jj'}(\Delta, \Delta') = E\{X_j(\Delta)\, X_{j'}(\Delta')\},$$

and in particular for $k=2$ we obtain for a complex function

$$X(\Delta)=X_1(\Delta)+iX_2(\Delta)$$

the expression
$$\sum_{p,q} h_p \bar{h}_q K(\Delta_p, \Delta_q)$$
with a complex kernel
$$K(\Delta, \Delta')=E\{X(\Delta)\,\bar{X}(\Delta')\}.$$

But, as on E_l so also on Ω, an *arbitrary* bi-distributive function $Q_2(\hat{\omega})$ on L_0 gives rise to a Gaussian process because the consistency relations (6.1.4) are then obviously fulfilled, and hence the following conclusion:

THEOREM 6.1.2. *For any admissible D, any positive-definite kernel is a covariance kernel, and in particular any Hilbert kernel is a covariance kernel.*

We now take in E_l: (α_p) a function $\phi(\alpha_p)$ which in the neighborhood of the origin $(\alpha)=(0)$ can be written, for some $t \geqq 1$, as a sum

$$\phi(\alpha_p)=1+P^r(\alpha_p)+R^r(\alpha_p), \tag{6.1.7}$$

where $P^r(\alpha_p)$ is a polynomial

$$P^r(\alpha_p)=\sum_{1\leqq n_1+\ldots+n_l\leqq r} a_{n_1\cdots n_l}(-2\pi i\alpha_1)^{n_1}\ldots(-2\pi i\alpha_l)^{n_l}, \tag{6.1.8}$$

and $R^r(\alpha_p)$ is of smaller order at the origin,

$$R^r(\alpha_p)=o(|\alpha|^r), \quad |\alpha|\to 0. \tag{6.1.9}$$

By taking the logarithm of (6.1.7) we can obviously write

$$\phi(\alpha_p)=\exp\left(P^r(\alpha_p)+S^r(\alpha_p)\right), \tag{6.1.10}$$

where again $S^r(\alpha_p)=o(|\alpha|^r)$, and (6.1.10) conversely implies (6.1.7). The representation (6.1.10) is unique. Otherwise we would have a relation

$$A(\alpha_p)=B(\alpha_p), \tag{6.1.11}$$

in which $B(\alpha_p)=o(|\alpha|^r)$ and $A(\alpha_p)$ is again a polynomial (6.1.8) and not $\equiv 0$. For some $s\leqq r$ we could then put

$$A(\alpha_p)=\sum_{\rho=s}^{r} C_\rho(\alpha_p),$$

where each $C_\rho(\alpha_p)$ is a *homogeneous* polynomial of degree s and $C_s(\alpha_p)\not\equiv 0$. However, if we introduce spherical coordinates $\alpha_p=\beta_p|\alpha|$, $\beta_1^2+\ldots+\beta_l^2=1$, and divide (6.1.11) by $|\alpha|^s$ and let $\alpha\to 0$, then we obtain $C_s(\beta_p)\equiv 0$, which would be contradictory.

If in a stochastic process each characteristic function is of the form

$$\phi_\lambda(\alpha_p)=\exp(P^r_\lambda(\alpha_p)+S^r_\lambda(\alpha_p))$$

for the same r, then (6.1.1) implies

$$P^r_\lambda(\alpha_p)+S^r_\lambda(\alpha_p)=P^r_\mu(g_{\mu\lambda}(\alpha))+S^r_\mu(g_{\mu\lambda}(\alpha)),$$

and because of the uniqueness just established this implies the relations

$$\left.\begin{aligned}P^r_\lambda(\alpha)=P^r_\mu(g_{\mu\lambda}(\alpha)),\\ S^r_\lambda(\alpha)=S^r_\mu(g_{\mu\lambda}(\alpha)),\end{aligned}\right\} \qquad (6.1.12)$$

separately. In fact, if $P^r_\mu(\beta)$ is a polynomial of degree r in the β's, then $P^r_\mu(g_{\mu\lambda}(\alpha))$ is such a polynomial in the α's, and $S^r_\mu(\beta)=o(|\beta|^r)$ as $\beta\to 0$ implies $S^r_\mu(g_{\mu\lambda}(\alpha))=o(|\alpha|^r)$ as $\alpha\to 0$.

As we have seen in section 3.5, if $r=2m$ then we have a representation (6.1.7) if and only if the distribution function $F(A)$ pertaining to $\phi(\alpha_p)$ has moments of order $\leqq 2m$, and we then have in fact

$$a_{n_1\ldots n_l}=\int_{E_l} x_1^{n_1}\ldots x_l^{n_l}\,dF(x_p).$$

For $r=2m+1$ the connection is not so direct, but in the next theorem we will denote the coefficients $a_{n_1\ldots n_l}$ as moments nevertheless. The first moments $a_{0\ldots1\ldots0}$ which are usually called *means* and denoted by $(m_1,\ldots,m_l)$ can always be 'removed', that is, made equal to zero, by envisaging the translated set function $F(A-m)$ instead of the original $F(A)$. Hence the following theorem:

THEOREM 6.1.3. (i) *In any Euclidean stochastic process, especially in any random function* $X(\Delta)$, *the means, if existing, can be made equal to zero in all distribution functions* $\{F_l(A)\}$ *simultaneously, and the moments of any one order have values connected with each other by relations* (6.1.12).

(ii) *If the process has (first and) second moments then there is another process (with the same consistency transformations* $g_{\mu\lambda}(\alpha)$) *which is Gaussian and has the same first and second moments as the given one.*

Part (ii) of the theorem is the most comprehensive precise version known to us of the proposition that if in a 'stochastic process' (however this term be defined) only first and second moments of joint distribution functions are being envisaged, then any conclusion that can be drawn for the particular case of Gaussian processes will also be valid for processes other than Gaussian.

Finally, if a stochastic process is Gauss–Poisson, then the following precise decomposition statement can be made even though the second moments need not exist any longer.

THEOREM 6.1.4. *If in a Gauss–Poisson process we put* (6.1.3) *in the form*

$$\phi_\lambda(\alpha_p)=\exp[\psi^G_\lambda(\alpha_p)+\psi^R_\lambda(\alpha_p)+i\psi^J_\lambda(\alpha_p)], \tag{6.1.13}$$

as in section 4.3, *then the Gaussian characteristic functions*

$$\exp[\psi^G_\lambda(\alpha_p)] \tag{6.1.14}$$

and the Bernoulli–Poisson functions

$$\exp[\psi^R_\lambda(\alpha_p)],\quad \exp[\psi^R_\lambda(\alpha_p)+i\psi^J_\lambda(\alpha_p)] \tag{6.1.15}$$

separately constitute a stochastic process each.

Proof. The relation

$$\psi^G_\lambda(\alpha)+\psi^R_\lambda(\alpha)+i\psi^J_\lambda(\alpha)=\psi^G_\mu(\beta)+\psi^R_\mu(\beta)+i\psi^J_\mu(\beta)$$

splits into the real and imaginary parts

$$\psi^G_\lambda(\alpha)+\psi^R_\lambda(\alpha)=\psi^G_\mu(\beta)+\psi^R_\mu(\beta), \tag{6.1.16}$$

$$\psi^J_\mu(\alpha)=\psi^J_\mu(\beta), \tag{6.1.17}$$

and (6.1.16) is the specific relation

$$\sum_{p,q=1}^{l} a^\lambda_{pq}\alpha_p\alpha_q+\int_{E'_l}(\sin(\alpha,x))^2\,dF_l(x)$$
$$=\sum_{r,s=1}^{m} a^\mu_{rs}\beta_r\beta_s+\int_{E'_m}(\sin(\beta,y))^2\,dF_m(y). \tag{6.1.18}$$

Now, the second integral is

$$\int_{E'_m}\left(\sin\left(\sum_{p=1}^{l}\alpha_p\eta_p\right)\right)^2 dF_m(y), \tag{6.1.19}$$

where
$$\eta_p=\sum_{q=1}^{m} c_{pq}y_q,$$

and the latter is a transformation from E'_m to E'_l. By using its inverse we can define a set function $F'_l(A)$ in E'_l such that (6.1.19) is identical with

$$\int_{E'_l}(\sin(\alpha,\eta))^2\,dF'_l(\eta),$$

but by the uniqueness assertion in theorem 3.4.2, relation (6.1.18) now splits into

$$\psi^G_\mu(\alpha)=\psi^G_\mu(\beta),$$

and also $\psi^R_\lambda(\alpha)=\psi^R_\mu(\beta)$. If we combine the latter with (6.1.17) we obtain

$$\psi^R_\lambda(\alpha)+i\psi^J_\lambda(\alpha)=\psi^R_\mu(\beta)+i\psi^J_\mu(\beta),$$

and this completes the proof of the theorem.

6.2. Convergence in probability and integration

Convergence in probability was introduced in definition 3.7.5, but for the present purposes another definition, known to be equivalent to it, will be more appropriate.

DEFINITION 6.2.1. On a probability space $\{\Omega; \mathscr{S}; P\}$ a sequence of random variables $x_n = f_n(\omega)$, each determined only a.e., is said to *converge* in *probability* (in measure) to a random variable $x = f(\omega)$, in symbols

$$x_n \to x \text{ (prob)},$$

if for each $\epsilon > 0$ we have

$$\lim_{n\to\infty} P(|f_n(\omega) - f(\omega)| > \epsilon) = 0, \tag{6.2.1}$$

and a sequence $\{x_n\}$ is a *Cauchy sequence* (prob) if

$$\lim_{m,\, n\to\infty} P(|f_m(\omega) - f_n(\omega)| > \epsilon) = 0 \tag{6.2.2}$$

for each $\epsilon > 0$.

The following properties of limits in probability will be taken as known:

LEMMA 6.2.1. *A sequence is convergent towards some limit element if (and only if) it is a Cauchy sequence, and the limit is unique a.e. A limit of limits of a set of random variables is again a limit of the set. If $x_n \to x$, $y_n \to y$ and α, β are real, then $\alpha x_n + \beta y_n \to \alpha x + \beta y$.*

THEOREM 6.2.1. *For a Euclidean stochastic process, if we are given a sequence $\{\hat{\omega}_n\}$ in $\hat{\Omega}$, then the corresponding sequence of random variables*

$$x_n = (\omega, \hat{\omega}_n) \tag{6.2.3}$$

on Ω is a Cauchy sequence (prob) if and only if for the characteristic functional we have

$$\lim_{m,\, n\to\infty} \phi(\alpha(\hat{\omega}_m - \hat{\omega}_n)) = 1 \tag{6.2.4}$$

for every real α, and the limit (prob) of (6.2.3) is $x_0 = (\omega, \omega_0)$ for $\hat{\omega}_0$ in $\hat{\Omega}$ if and only if

$$\lim_{n\to\infty} (\alpha(\omega_n - \omega_0)) = 1. \tag{6.2.5}$$

Proof. It will suffice to prove the first part of the theorem. If $c(x)$ is bounded continuous in $(-\infty, \infty)$, then, as is easily seen, for any Cauchy sequence we have

$$\lim_{m,\, n\to\infty} \int_{\Omega} c(f_m(\omega) - f_n(\omega))\, dP(\omega) = c(0),$$

and on putting $x_n = f_n(\omega)$ and $c(x) = e^{-2\pi i \alpha x}$ we obtain (6.2.4). Conversely if, for instance, in

$$e^{-\pi x^2} = \int_{-\infty}^{\infty} e^{-\pi\alpha^2 - 2\pi i \alpha x}\, d\alpha$$

we put $x=(\omega, \hat{\omega}_m)-(\omega, \hat{\omega}_n)$ and integrate over Ω we obtain

$$E\{e^{-\pi |x_m-x_n|^2}\}=\int_{-\infty}^{\infty} e^{-\pi\alpha^2}\phi(\alpha(\hat{\omega}_m-\hat{\omega}_n))\,d\alpha,$$

so that (6.2.4) implies

$$\lim_{m,\,n\to\infty} E\{e^{-\pi |x_m-x_n|^2}\}=1.$$

But this implies (6.2.2), as claimed.

THEOREM 6.2.2. *Let $\hat{\Omega}$ be an everywhere dense part of a larger vector space $\hat{\Omega}^*$: $(\hat{\omega}^*)$ on which there is given a topology in which $\alpha\hat{\omega}_1^*+\beta\hat{\omega}_2^*$ is continuous in the four quantities shown, and let $\hat{\Omega}^*$ be complete in the sense that a sequence $\{\hat{\omega}_n^*\}$ has a limit point if and only if we have*

$$\lim_{m,\,n\to\infty} (\hat{\omega}_m^*-\hat{\omega}_n^*)=0 \quad (\textit{in topology}). \tag{6.2.6}$$

If a characteristic functional $\phi(\hat{\omega})$ on $\hat{\Omega}$ is continuous in this topology, then there exists a distributive operation $(\omega, \hat{\omega}^*)$ from $\hat{\Omega}^*$ to the vector space of random variables on Ω, which for $\hat{\omega}^*=\hat{\omega}$ reduces to the original expression $(\omega, \hat{\omega})$ and which is such that

$$\hat{\omega}_n^*\to\hat{\omega}_0^* \quad (\textit{in topology}) \tag{6.2.7}$$

implies

$$(\omega, \hat{\omega}_n^*)\to(\omega, \hat{\omega}_0^*) \ \text{(prob)}. \tag{6.2.8}$$

Proof. Every $\hat{\omega}^*$ is the limit of a sequence $\{\hat{\omega}_n\}$ in $\hat{\Omega}$, and by theorem 6.2.1 the corresponding sequence (6.2.3) has a limit (prob). Also, by a property stated in lemma 6.2.1, this limit depends only on $\hat{\omega}^*$ itself and not on the approximating sequence chosen, and thus may be denoted by $(\omega, \hat{\omega}^*)$. The properties claimed for the latter entity may now be obtained by full use of lemma 6.2.1.

We now take an admissible family $D=\{\Delta\}$ on a space T, and denote by $D^*=\{\Delta^*\}$ the smallest σ-field of sets on T containing it, and we assume that on D^* there is given a *Lebesgue measure* $v(\Delta^*)$ which for the elements of D is both *finite and* $\neq 0$.

For $\rho>0$, the measurable functions $h(t)$ for which

$$\delta(h)\equiv\begin{cases}\left(\displaystyle\int_T |h(t)|^\rho\,dv_t\right)^{1/\rho} & \text{if} \quad 1\leqq\rho<\infty \\ \displaystyle\int_T |h(t)|^\rho\,dv_t & \text{if} \quad 0<\rho\leqq 1\end{cases} \tag{6.2.9}$$

is finite are determined almost everywhere, and they constitute a vector space $L_\rho\supset L_0$. Also, $\delta(h_1-h_2)$ has the properties of a metric on L_ρ for which the vector operations are continuous and the space is complete (even for $0<\rho\leqq 1$, which is rarely so stated); and for $\hat{\Omega}=L_0$ the space L_ρ is a suitable extension $\hat{\Omega}^*$. Hence the following definition and theorem:

DEFINITION 6.2.2. We say that a random function $X(\Delta)$ is *L_ρ-finite* if $\phi(h)$ on L_0 is continuous in the L_ρ-metric.

Note that if $v(T)<\infty$ (thus, for instance, if T is a *bounded* interval), then any function which is L_ρ-finite is also $L_{\rho'}$-finite for $\rho'>\rho$.

THEOREM 6.2.3. *If $X(\Delta)$ on D is L_ρ-finite for some $\rho>0$ then, by limits in probability, the integral*

$$\int_T h(t)\,dX(t) \tag{6.2.10}$$

can be extended from h in L_0 to h in L_ρ, and the 'extended' integral is continuous (prob) relative to the L_ρ-metric.

In particular, by putting $h(t)=\omega_{\Delta*}(t)$, $X(\Delta)$ can be extended to all sets Δ^* for which $v(\Delta^*)<\infty$, and the extension is σ-additive (prob).

If $\phi(h)$ is of the form
$$\Phi\left(\int_T |h(t)|^\rho\,dv_t\right), \tag{6.2.11}$$

where $\Phi(\xi)$ is defined and continuous in $0\leqq\xi<\infty$, then it is obviously L_ρ-finite, and therefore the two processes

$$\phi(h)=\exp\left(-\int_T |h(t)|^2\,dv_t\right), \tag{6.2.12}$$

$$\phi(h)=\exp\left(-\left(\int_T |h(t)|^2\,dv_t\right)^{\frac{1}{2}\rho}\right) \tag{6.2.13}$$

are both L_2-finite. But of

$$\phi(h)=\exp\left(-\int_T |h(t)|^\rho\,dv_t\right), \tag{6.2.14}$$

we can only say that it is L_ρ-finite even though it looks very similar to (6.2.13). But (6.2.14) is a homogeneous Lévy process (see definition 5.5.4) just as (6.2.12) is homogeneous Gaussian; (6.2.13), however, is not homogeneous, although it is subordinate to (6.2.12).

Remark. It is easily seen that every homogeneous symmetric Gaussian $X(\Delta)$ must be of the form (6.2.12), if $X(\Delta)$ is 'scalar'. But if it is a vector $[X^1(\Delta),\ldots,X^k(\Delta)]$, then the general form of its characteristic functional is

$$\phi(h^1,\ldots,h^k)=\exp\left(-\int_T \sum_{p,q=1}^{k} h^p(t)\,h^q(t)\,dv_{pq}(t)\right), \tag{6.2.15}$$

where $v_{pq}(\Delta)$ is a (nonpositive) real additive function on D such that $\sum_{p,q} h^p h^q v_{pq}(\Delta)$ is positive semidefinite for each Δ.

In general, for $k=2$, we can introduce complex quantities $X(\Delta)$, $h(t)$ as in remark of section 5.5. Now, if the sequence

$$\int_T [h'_n\, dX'(t) + h''_n dX''(t)]$$

is convergent (prob) for *every* sequence $\{h'_n, h''_n\}$ which is convergent in L_ρ-metric, then so is also the sequence

$$\int_T [h''_n dX'(t) - h'_n dX''(t)],$$

hence the following statement:

THEOREM 6.2.4. *If a complex random function* $X(\Delta)$ *is* L_ρ*-finite, then the integrals*

$$\int_T \overline{h(t)}\, dX(t) \tag{6.2.16}$$

and $\int_T h(t)\, \overline{dX(t)}$ *have the completeness and closure properties stated in theorem* 6.2.3.

DEFINITION 6.2.3. We will call a complex random function $X(\Delta)$ a *Wiener process* if its characteristic functional is

$$\phi(h) = \exp\left(-\int_T |h(t)|^2 dv\right). \tag{6.2.17}$$

If written in real components, a Wiener process is a Gaussian vector $[X^1(\Delta), X^2(\Delta)]$ with characteristic functional

$$\phi(h^1, h^2) = \exp\left(-\int_T ((h^1)^2 + (h^2)^2)\, dv\right), \tag{6.2.18}$$

and this is a special case of (6.2.15) for $k=2$.

Henceforth all functions $X(\Delta)$ will be complex.

6.3. Convergence in norm

If for given $p \geqq 1$ we consider on the given probability space only those random variables $x = f(\omega)$ for which the $L_p(\Omega)$-norm

$$\|f\| = \left(\int_\Omega |f(\omega)|^p\, dP(\omega)\right)^{1/p} \tag{6.3.1}$$

is finite, then we can envisage convergence in norm, $\|f_n - f\| \to 0$, or

$$f_n \to f(L_p), \tag{6.3.2}$$

and we note the following statement which is easily proven:

LEMMA 6.3.1. *If* $f_n \to f(L_p)$ *then* $f_n \to f$ (prob); *but not necessarily conversely.*

DEFINITION 6.3.1. We say that $X(\Delta)$ is *$L_{\rho,p}$-bounded* if $X(\Delta)$ is in $L_p(\Omega)$, meaning that

$$Y=\int_T \overline{h(t)}\,dX(t) \tag{6.3.3}$$

is in $L_p(\Omega)$ for $h\in L_0$, and if there is a constant M such that

$$\int |h(t)|^\rho\,dv \leqq 1$$

implies

$$E\{|Y|^p\}\leqq M. \tag{6.3.4}$$

Lemma 6.3.1 implies as follows:

THEOREM 6.3.1. *For any $\rho>0$, if $X(\Delta)$ is $L_{\rho,p}$-bounded for some $p\geqq 1$, it is also L_ρ-finite.*

If a process is of the form (6.2.12) or (6.2.13) or (6.2.14) then the (one-dimensional) characteristic function of the random variable (6.3.3) is of the form

$$\exp\left(-\alpha^2\int_T |h|^2\,dv\right),\quad \exp\left(-|\alpha|^\rho\left(\int_T |h|^2\,dv\right)^{\frac{1}{2}\rho}\right),$$
$$\exp\left(-|\alpha|^\rho\int_T |h|^\rho\,dv\right)$$

respectively, and with the aid of theorem 3.5.3 we may state as follows:

THEOREM 6.3.2. *A Wiener process is $L_{2,p}$-bounded for every $p\geqq 1$ and is in particular $L_{2,2}$-bounded.*

The process (6.2.13), *which is subordinate to a Wiener process, is $L_{2,p}$-bounded for $p<\rho$; and the homogeneous Lévy process* (6.2.14) *is $L_{\rho,p}$-bounded for $p<\rho$.*

We now start out not with a set function $X(\Delta)$ but with a (random) point function $x(t)$ on T with values in $L_p(\Omega)$, and we assume that this vector-valued function $x(t)$ is (strongly) measurable in $v(\Delta^*)$ as in section 1.3, and that for some $\sigma<1$ the integral

$$|||x|||=\left(\int_T \|x(t)\|^\sigma\,dv\right)^{1/\sigma} \tag{6.3.5}$$

is finite, thus being a Banach norm itself. The number σ need have no relation to the number p in (6.3.1) and in fact the norm (6.3.5) could be introduced for x in some Banach space B not necessarily an $L_p(\Omega)$, and in order to emphasize this we will denote the Banach space of functions $\{x(t)\}$ with norm (6.3.5) by $L_\sigma(B)$ even for a given p. We will also consider the norm (6.3.5) for $\sigma=\infty$ defining it then by

$$|||x|||=\text{essential upper bound of}\ \|x(t)\|,\quad t\in T$$

so that we can introduce the quotient

$$\sigma = \frac{\rho}{\rho - 1} \tag{6.3.6}$$

for all $\rho \geqq 1$, $\rho = 1$ included.

By the theory of integration of vector-valued functions, if

$$h(t) \in L_\rho(T) \quad \text{and} \quad x(\cdot) \in L_\sigma(B),$$

then the integral

$$y = \int_T \overline{h(t)}\, x(t)\, dv(t) \tag{6.3.7}$$

exists (and is an element of $B \equiv L_p(\Omega)$ again), and, in fact, by Holder's inequality we have

$$\| y \| \leqq \left(\int_T | h(t) |^\rho \, dv \right)^{1/\rho} \cdot |||x|||. \tag{6.3.8}$$

In particular, $h(t) \equiv \omega_\Delta(t) \epsilon L_\rho(T)$, so that the integral

$$\int_T \omega_\Delta(t)\, x(t)\, dv \equiv \int_\Delta x(t)\, dv \equiv X(\Delta) \tag{6.3.9}$$

exists and is an element of $L_p(\Omega)$, and as a function on D it is additive and in fact σ-additive (L_p). For $h \in L_0$, the integral (6.3.7) can now be equated with the integral (6.3.3) and hence the following statement:

THEOREM 6.3.3. *If $X(\Delta)$ is an indefinite integral* (6.3.9), *then it is $L_{\rho, p}$-bounded.*

This theorem does not have a converse; and we will see that even the Wiener process, which is the most 'bounded' process known, is *not* an indefinite integral if the variable t is continuous.

DEFINITION 6.3.2. We say that the measure $v(\Delta)$ is *L_2-atomistic* if for every $\epsilon > 0$ there is a $\delta > 0$ such that $\sum_m v(\Delta_m) \leqq \delta$ implies $\sum_m v(\Delta_m)^{\frac{1}{2}} \leqq \epsilon$ for any sum $\Delta_1 + \dots + \Delta_l$.

THEOREM 6.3.4. *If $X(\Delta)$ is a Wiener process, then $X(\Delta)$ is the integral* (6.3.9) *of a function $x(t)$ with values in $L_2(\Omega)$ if and only if $v(\Delta)$ is L_2-atomistic.*

Proof. For (6.3.3) we have

$$E\{| Y |^2\} = c^2 \int_T | h(t) |^2 \, dv$$

for a constant $c > 0$, and hence

$$\sum_{m=1}^{l} \| X(\Delta_m) \| \equiv \sum_{m=1}^{l} (E\{| X(\Delta_m) |^2\})^{\frac{1}{2}} \equiv c \sum_{m=1}^{l} | v(\Delta_m) |^{\frac{1}{2}},$$

so that $X(\Delta)$ is 'absolutely continuous' in $L_2(\Omega)$-norm if and only if $v(\Delta)$ is L_2-atomistic, as claimed.

Ordinary measure on $a<t<b$ is not L_2-atomistic, and thus 'ordinary' Wiener process is not an indefinite integral of a function $x(t)$ with values x in $L_2(\Omega)$.

If we apply definition 6.1.1 to a complex function $X(\Delta)$, then it is a Gaussian process if its characteristic function is of the form

$$\exp(iQ_1(h)-Q_2(h)), \tag{6.3.10}$$

where $Q_1(h)$ is real distributive and $Q_2(h)$ is non-negative Hermitian bi-distributive on the complex L_0.

THEOREM 6.3.5. *A Gaussian process $X(\Delta)$ is $L_{2,2}$-bounded if and only if $Q_1(h)$ and $Q_2(h)$ are both bounded for*

$$\int_T |h(t)|^2 dv \leqq 1, \tag{6.3.11}$$

that is, if, and only if, on the complex $L_2(T)$, $Q_1(h)$ is a real linear functional and $Q_2(h)=(Ah,h)$ where Ah is a non-negative bounded self-adjoint operator.

Proof. The characteristic function of (6.3.3) is

$$\exp(i\alpha Q_1(h)-\alpha^2 Q_2(h)),$$

and thus, except for a constant factor,

$$E\{|Y|\}^2 = Q_2(h)+\tfrac{1}{2}(Q_1(h))^2.$$

But this is bounded in the unit sphere (6.3.11) if, and only if, $Q_1(h)$, $Q_2(h)$ are bounded separately, as claimed.

6.4. Expansion in series and integrals

We assume that there exists on $\{D^*; v(\Delta^*)\}$ a *complete* orthonormal system

$$\{g_n(t)\}, \quad n=1,2,3,\ldots, \tag{6.4.1}$$

with

$$\int_T g_m(t)\overline{g_n(t)}\,dv=\delta_{mn}. \tag{6.4.2}$$

For any $h(t)\in L_2(T)$ we can introduce the expansion

$$h(t)\sim\sum_1^\infty \gamma_m g_m(t), \qquad \gamma_m=\int_T h(t)\overline{g_n(t)}\,dv, \tag{6.4.3}$$

and we have

$$h(t)=\lim_{n\to\infty}\sum_{m=1}^n \gamma_m g_m(t)\ (L_2), \tag{6.4.4}$$

by the Riesz–Fischer theorem.

If for some $\rho\geqq 1$ the functions $g_n(t)$ belong to both L_ρ and $L_{\rho(\rho-1)}$, then the series (6.4.3) can also be set up for $h \in L_\rho(T)$, and it then frequently happens that there exists a matrix of real numbers

$$\{\lambda_{nm}\}, \quad m, n = 1, 2, 3, \ldots,$$

in which for each n only finitely many are $\neq 0$ such that

$$h(t) = \lim_{n \to \infty} \sum_{m=1}^{\infty} \lambda_{nm} \gamma_m g_m(t) \, (L_\rho). \tag{6.4.5}$$

For instance, for ordinary periodic functions $h(t)$ in $-\frac{1}{2} \leqq t < \frac{1}{2}$, if we put

$$h(t) \sim \sum_{-\infty}^{\infty} \gamma_m e^{2\pi i m t}, \quad \gamma_m = \int_{\frac{1}{2}}^{\frac{1}{2}} h(t)\, e^{2\pi i m t} \, dt, \tag{6.4.6}$$

then, as we know, we have

$$h(t) = \lim_{n \to \infty} \sum_{-n}^{n} \left(1 - \frac{|m|}{n}\right) \gamma_m e^{2\pi i m t} \, (L_\rho), \tag{6.4.7}$$

for $h(t) \in L, \rho \geqq 1$; and, in fact, by a renowned theorem of Marcel Riesz we have

$$h(t) = \lim_{n \to \infty} \sum_{-n}^{n} \gamma_m e^{2\pi i m t} \, (L_\rho), \tag{6.4.8}$$

not only for $\rho = 2$ but also for $\rho > 1$, although this is of no particular consequence to us.

Now, if we substitute all these expansions into our integral (6.3.3) we obtain the following theorem:

THEOREM 6.4.1. (i) *If for an L_ρ-finite $X(\Delta)$ we introduce the (random-valued) Fourier coefficients*

$$Y_m = \int_T \overline{g_m(t)} \, dX(t), \tag{6.4.9}$$

then we have

$$\frac{dX(\Delta)}{d\Delta} \sim \sum_{m=1}^{\infty} Y_m g_m(t) \tag{6.4.10}$$

in the sense that for $h \in L_\rho$ we have

$$\int_T \overline{h(t)} \, dX(t) = \lim_{n \to \infty} \sum_{m=1}^{\infty} \lambda_{nm} \bar{\gamma}_m Y_m \text{ (prob)}, \tag{6.4.11}$$

and

$$= \lim_{n \to \infty} \sum_{m=1}^{n} \bar{\gamma}_m Y_m \text{ (prob)} \tag{6.4.12}$$

for $h \in L_2$.

(ii) *In particular, if $X(\Delta)$ is periodic on $-\frac{1}{2} \leqq t < \frac{1}{2}$, then we have*

$$\frac{dX(\Delta)}{d\Delta} \sim \sum_{-\infty}^{\infty} Y_m e^{2\pi i m t}, \quad Y_m = \int_{-\frac{1}{2}}^{\frac{1}{2}} e^{-2\pi i m t} \, dX(t) \tag{6.4.13}$$

in the sense that

$$\int_{-\frac{1}{2}}^{\frac{1}{2}} \overline{h(t)}\, dX(t) = \lim_{n\to\infty} \sum_{m=-n}^{n} \left(1 - \frac{|m|}{n}\right) \bar{\gamma}_m Y_m \text{ (prob)} \quad (6.4.14)$$

for $\rho \geqq 1$, *and*

$$= \lim_{n\to\infty} \sum_{m=-n}^{n} \bar{\gamma}_m Y_m \text{ (prob)} \quad (6.4.15)$$

for $\rho > 1$.

(iii) *If, more precisely,* $X(\Delta)$ *is* $L_{\rho, p}$*-bounded, then the limits* (6.4.11), (6.4.12), (6.4.14) and (6.4.15) *exist not only in probability but also in* $L_p(\Omega)$*-norm.*

We are now viewing the sequence of random variables $\{Y_m\}$, as given by (6.4.9), as a random-valued function on the 'discrete' set $T' = \{m\}$, and we assign to each point m the measure 1. The analogue to the previous vector space L_0 will be denoted by L'_0, and it consists of sequences $\gamma = \{\gamma_m\}$ in which only finitely many components are $\neq 0$, and the space L'_ρ, which is meant to be the analogue to L_ρ, is determined by the norm

$$\left(\sum_{m=1}^{\infty} |\gamma_m|^\rho\right)^{1/\rho}$$

for $\rho \geqq 1$.

THEOREM 6.4.2. (i) *If* $\{X(\Delta)\}$ *is* L_2*-finite, then* $\{Y_m\}$ *is* L_2*-finite; and conversely. Similarly for* $L_{2,p}$*-boundedness.*

(ii) *If* $\{X(\Delta)\}$ *is* $L_{2,2}$*-bounded Gaussian, then* $\{Y_m\}$ *is* $L_{2,2}$*-bounded Gaussian; and conversely.*

(iii) *If* $\{X(\Delta)\}$ *is a Wiener process, then* $\{Y_m\}$ *is a Wiener process; and conversely. (However, if a non-Gaussian process* $\{X(\Delta)\}$ *is homogeneous, then* $\{Y_m\}$ *need not be homogeneous; nor conversely.)*

Proof. If $\{X(\Delta)\}$ is L_2-finite, and if with $\{\gamma_m\}$ *in* L'_0 we form

$$h(t) = \sum_m \gamma_m g_m(t), \quad (6.4.16)$$

then we obtain

$$\int_T \overline{h(t)}\, dX(t) = \sum_m \bar{\gamma}_m Y_m, \quad (6.4.17)$$

and by Parseval's equation we have

$$\int_T |h(t)|^2\, dv = \sum_m |\gamma_m|^2. \quad (6.4.18)$$

If now we take a sequence of elements $\{\gamma_m^r\}$ in L'_0, $r = 1, 2, \ldots$, and if we apply these formulas to the differences $\gamma_m = \gamma_m^r - \gamma_m^s$, and if we let $r, s \to \infty$, then we conclude that $\{Y_m\}$ is also L'_2-finite.

Conversely, if $\{Y_m\}$ is L'_2-finite, and if for $h \in L_0$ we introduce the expansion (6.4.3), then (6.4.12) implies (6.4.17), and if we now take a sequence of elements $\{h^r\}$ in L_0 and apply this and (6.4.18) to the differences $h = h^r - h^s$, $r, s \to \infty$, then $\{X(\Delta)\}$ turns out to be L_2-finite. Also, by the same reasoning, if either process is $L_{2,p}$-bounded then so is the other.

Next, if $\{X(\Delta)\}$ is Gaussian, then its characteristic functional

$$E\left\{\exp\left(i\int_T \overline{h(t)}\, dX(t)\right)\right\} \tag{6.4.19}$$

has the form (6.3.10). But by (6.4.15), (6.4.19) is identical with

$$E\left\{\exp\left(i\sum_m \bar{\gamma}_m Y_m\right)\right\}, \tag{6.4.20}$$

which is the characteristic functional on $\{Y_m\}$, and on the other hand the substitution (6.4.15) also transforms (6.3.10) into

$$\exp\,(iQ'(\gamma) - Q''(\gamma)), \tag{6.4.21}$$

where again $Q'(\gamma)$ is real continuous on L_2' and $Q''(\gamma)$ is non-negative bounded self-adjoint there. Thus $\{Y_m\}$ is also Gaussian. Finally, the identity (6.4.18) shows that if either process is a Wiener process then so is the other.

Remark. The above reasoning also gives the following conclusion. If $\{X(\Delta)\}$ is subordinate to an $L_{2,2}$-bounded Gaussian process, thus having a characteristic functional of the form

$$\Phi(iQ_1(h) - Q_2(h)), \tag{6.4.22}$$

with 'bounded' $Q_1(h)$ and $Q_2(h)$, then $\{Y_m\}$ is of the same kind and has 'the same' characteristic functional; and conversely.

We are now turning to Fourier integrals, and the following theorem, although insufficient for important applications, is a centerpiece of the L_2-theory:

THEOREM 6.4.3. *If $X(\Delta)$ is defined and $L_{2,2}$-bounded on*

$$E_1\colon (-\infty < t < \infty),$$

then there exists a (unique) other process $Y(\Delta)$, likewise defined and $L_{2,2}$-bounded on E_1, such that we have

$$\int_{-\infty}^{\infty} \overline{h(r)}\, dX(r) = \int_{-\infty}^{\infty} \overline{g(s)}\, dY(s) \tag{6.4.23}$$

for any $h(r), g(s) \in L_2(E_1)$ which are Plancherel transforms

$$h(r) \sim \int_{-\infty}^{\infty} e^{2\pi i r s} g(s)\, ds, \quad g(s) \sim \int_{-\infty}^{\infty} e^{-2\pi i s r} h(r)\, dr \tag{6.4.24}$$

of each other.

In particular, by putting $g(s) = \omega_\Delta(s)$, *we have*

$$Y(\Delta_{\alpha\beta}) = \int_{-\infty}^{\infty} \left(\int_{\alpha}^{\beta} e^{-2\pi i s r}\, ds \right) dX(r) \equiv \int_{-\infty}^{\infty} \frac{e^{-2\pi i r\beta} - e^{-2\pi i r\alpha}}{-2\pi i}\, dX(r) \tag{6.4.25}$$

and inversely (or 'dually'),

$$X(\Delta_{\alpha\beta}) = \int_{-\infty}^{\infty} \frac{e^{2\pi i s\beta} - e^{2\pi i s\alpha}}{2\pi i}\, dY(s). \tag{6.4.26}$$

Also, if either process is Gaussian or Wiener, then so is the other, and if the characteristic functional of $X(\Delta)$ *has the form* (6.4.22), *then the characteristic functional of* $Y(\Delta)$ *arises by insertion of the first integral* (6.4.24) *into it.*

Proof. If $X(\Delta)$ is $L_{2,2}$-bounded on E_1, then

$$Y_h \equiv \int_{-\infty}^{\infty} \overline{h(r)}\, dX(r) \tag{6.4.27}$$

is a bounded linear operation from $L_2(E_1)$ to $L_2(\Omega)$, and it is readily seen that any such operation can be so represented by a function $X(\Delta)$, and uniquely so. Now the Plancherel transformation (6.4.24) is a unitary mapping of $L_2(E_1)$ into itself, and such a mapping carries a bounded linear operation from $L_2(E_1)$ to $L_2(\Omega)$ into another such one, whence the theorem.

6.5. Stationarity and orthogonality

Joint assumption. In the present section every process occurring is $L_{2,2}$-bounded on its entire space of existence.

DEFINITION 6.5.1. We say that $X(\Delta)$ on $(-\infty, \infty)$ is K-stationary ('K' for Khintchine), if for any two finite intervals Δ_m: $\alpha_m < r \leqq \beta_m$, $m = 1, 2$, and all its translates Δ_m^t: $\alpha_m + t < r \leqq \beta_m + t$, we have

$$E\{X(\Delta_1^t)\,.\,\overline{X(\Delta_2^t)}\} = E\{X(\Delta_1)\,.\,\overline{X(\Delta_2)}\} \tag{6.5.1}$$

LEMMA 6.5.1. $X(\Delta)$ *is* K*-stationary if and only if the bilinear functional*

$$P(h^1(\cdot), h^2(\cdot)) = E\left\{ \int_{-\infty}^{\infty} \overline{h^1(r)}\, dX(r)\,.\int_{-\infty}^{\infty} h^2(s)\, \overline{dX(x)} \right\}, \tag{6.5.2}$$

$h^1, h^2 \in L_2$, *is commutative with translations,*

$$P(h^1(r+t), \quad h^2(s+t)) \equiv P(h^1(r), \quad h^2(s)), \tag{6.5.3}$$

$-\infty < t < \infty$. *Also, if* $X(\Delta)$ *happens to be the indefinite integral* (6.3.9) *of an* L_2*-continuous function* $x(t)$ *then* $X(\Delta)$ *is* K*-stationary if and only if for the covariance function*

$$R(r, s) = E\{x(r)\,\overline{x(s)}\}, \tag{6.5.4}$$

we have $$R(r+t, s+t) = R(r, s),$$

so that we have $$R(r, s) = R(r-s) \tag{6.5.5}$$

for a (continuous) function $R(t)$.

The proof of the first part follows easily if we approximate to h^1, h^2 by step-function, and of the second part if we substitute the integral (6.3.9) into condition (6.5.1).

We also note without discussion (because it will be of no consequence to us) that if a K-stationary $X(\Delta)$ is an integral (6.3.9) of a function $x(t)$ which is integrable in $L_2(\Omega)$-norm, then, after adjustment on a t-set of measure zero, $x(t)$ is continuous in $L_2(\Omega)$-norm as envisaged in the second half of lemma 6.5.1.

DEFINITION 6.5.2. We call $Y(\Delta)$ on $(-\infty, \infty)$ *orthogonal* if we have

$$E\{Y(\Delta_1) . \overline{Y(\Delta_2)}\} = 0 \tag{6.5.6}$$

for any two intervals which are disjoint, $\Delta_1 \cap \Delta_2 = 0$.

We call it *orthonormal* if, furthermore,

$$E\{| Y(\Delta) |^2\} = c \, | \Delta |, \tag{6.5.7}$$

where $(\Delta) = \beta - \alpha$ is the length of the interval, and $c > 0$ is constant.

Remark 1. Requirements (6.5.6) and (6.5.7) can be contracted into

$$E\{Y(\Delta_1) . \overline{Y}(\Delta_2)\} = c \, | \Delta_2 \cap \Delta_2 |. \tag{6.5.8}$$

Also, $Y(\Delta)$ is orthonormal if and only if it is both orthogonal and K-stationary.

THEOREM 6.5.1. *If* $X(\Delta)$ *and* $Y(\Delta)$ *are transforms as in theorem* 6.4.3, *then either of them is K-stationary if and only if the other is orthogonal.*

In particular, therefore, if either of them is orthonormal then so is also the other.

Remark 2. Note that 'symmetry', Gaussian character and orthonormality are each 'invariant' under Fourier transformation, and a Wiener process is all three at the same time.

Proof. Assume first that $Y(\Delta)$ is orthogonal. For $\Delta_1 \cap \Delta_2 = 0$, we then have

$$E\{| Y(\Delta_1 + \Delta_2) |^2\} = E\{| Y(\Delta_1) |^2\} + E\{| Y(\Delta_2) |^2\},$$

and thus if we introduce

$$\Gamma(\Delta) = E\{| Y(\Delta) |^2\}, \tag{6.5.9}$$

then this is a non-negative additive interval function which, however, is defined for finite intervals only. It is important to note for applications in other contexts that thus far we have only used the assumptions that each $Y(\Delta)$ is an element of $L_2(\Omega)$, and that the orthogonality property (6.5.6) is available. However, from full $L_{2,2}$-boundedness we deduce

$$\Gamma(\Delta)=E\left\{\left|\int_{-\infty}^{\infty}\omega_{\Delta}(r)\,dY(r)\right|^2\right\}\leqq M\int_{-\infty}^{\infty}|\,\omega_{\Delta}(r)\,|^2\,dr=M\,.\,|\,\Delta\,|,$$

and by Lebesgue theory there exists now a bounded measurable function $\chi(\alpha)$,

$$0\leqq\chi(\alpha)\leqq M, \tag{6.5.10}$$

such that

$$\Gamma(\Delta_{r,s})=\int_r^s\chi(\alpha)\,d\alpha. \tag{6.5.11}$$

We now introduce the bilinear functional

$$Q(g^1(\cdot),g(\cdot))=E\left\{\int_{-\infty}^{\infty}\overline{g^1(r)}\,dY(r)\,.\int_{-\infty}^{\infty}g^2(s)\,d\overline{Y}(s)\right\}, \tag{6.5.12}$$

$g^1,g^2\in L_2$, and if g^1 and g^2 are both step functions on the same sum of intervals $\Delta_1+\ldots+\Delta_l$, and if g^m_j are the values of $g^m(r)$ on Δ_j; $m=1,2$; $j=1,\ldots,l$; then (6.5.12) has the value

$$\sum_{j=1}^{l}\overline{g^1_j}g^2_j\Gamma(\Delta_j).$$

Now, this expression is nothing but the integral

$$\int_{-\infty}^{\infty}\overline{g^1}(\alpha)\,g^2(\alpha)\,\chi(\alpha)\,d\alpha, \tag{6.5.13}$$

and by $L_{2,2}$-boundedness we obtain that this is the value of (6.5.12) for g^1,g^2 in general. From this we conclude that (6.5.12) has the invariance property

$$Q(g^1(r),g^2(s))\equiv Q(g^1(r)\,e^{2\pi irt},g^2(r)\,e^{2\pi irt}) \tag{6.5.14}$$

for all t; and if we introduce the Plancherel transforms (6.4.22) for both functions g^1, g^2, then (6.5.14) goes over into the relation (6.5.3), meaning that $X(\Delta)$ is K-stationary as claimed.

Conversely, assume that $X(\Delta)$ is K-stationary. Since, for $h^1=h^2=h$, $P(h,h)$ is real and 'bounded', it follows from Hilbert space theory that there exists a bounded self-adjoint operator Sh such that

$$P(h^1,h^2)=(h^2,Sh^1), \tag{6.5.15}$$

where the expression on the right is an 'inner product'. If we denote by U^th the operation which carries $h(\cdot)$ into $h(\cdot+t)$, then the stationarity assumption (6.5.3) can be stated as

$$(h^2,Sh^1)=(U^th^2,SU^th^1),$$

but the right side is also $(h^2, (U^t)^* SU^t h^1)$, so that we have

$$S = (U^t)^* SU^t.$$

Thus, S is commutative with translations, and since S is bounded, our lemma 4.6.1 becomes applicable. Therefore, if we denote the Plancherel transforms of g^1, g^2 by h^1, h^2, then $P(h^1, h^2)$ has the value (6.5.13), where $\chi(\alpha)$ is a bounded measurable function associable with the operator S. However, $Q(g^1, g^2) = P(h^1, h^2)$, and thus $Q(g^1, g^2)$ is expressible as an integral (6.5.13), and if we put $g^m(\alpha) = \omega_{\Delta_m}(\alpha)$, $m = 1, 2$, then (6.5.13) has the value 0 for $\Delta_1 \cap \Delta_2 = 0$, which proves (6.5.6), as claimed.

Our lemma 4.6.1 has an analogue, more or less, in every situation in which a Plancherel duality theorem is available, and theorem 6.5.1 has then also an analogue of some kind, in 'noncommutative' cases as well. The following statement which will not be further discussed is typical of them all:

THEOREM 6.5.2. *If $\{X(\Delta)\}$ is periodic on $-\frac{1}{2} \leqq t < \frac{1}{2}$, and if*

$$Y_m = \int_{-\frac{1}{2}}^{\frac{1}{2}} e^{-2\pi i m t} \, dX(t) \tag{6.5.16}$$

is the 'dual' process, then $X(\Delta)$ is stationary according to (the literally the same) definition 6.5.1, if and only if $\{Y_m\}$ is orthogonal in the sense that

$$E\{Y_m \overline{Y}_n\} = 0, \quad m \neq n;$$

and $X(\Delta)$ is orthogonal according to (the literally the same) definition 6.5.2 if and only if $\{Y_m\}$ is stationary in the sense that

$$R(m, n) = E\{Y_m \overline{Y}_n\}$$

is a function of $n - m$ only, $R(m, n) = R(m - n)$.

Also, $\{X(\Delta)\}$ is orthonormal according to definition 6.5.2 if and only if $\{Y_m\}$ is orthonormal in the familiar sense that

$$E\{Y_m \overline{Y}_n\} = c\delta_{mn},$$

(usually with $c = 1$).

6.6. Further statements

Theorem 6.5.1 is not the familiar proposition due to Khintchine and others. The latter does not introduce $L_{2,2}$-boundedness or feature duality so prominently, but, in an indirect fashion it introduces $L_{1,2}$-boundedness, and it is crucially based on our theorem 3.2.3 about positive-definite functions which it applies by way of the following proposition:

THEOREM 6.6.1. *If a random point function $x(t)$ is continuous in $L_2(\Omega)$-norm, and if we have*

$$E\{x(r)\overline{x(s)}\}=R(r-s), \tag{6.6.1}$$

then we can put

$$R(r-s)=\int_{-\infty}^{\infty} e^{2\pi i(r-s)\alpha}\,d\Gamma(\alpha), \tag{6.6.2}$$

where

$$\Gamma(A)\geqq 0,\quad \Gamma(E_1)\equiv R(0)<\infty. \tag{6.6.3}$$

For any $h^1(\cdot), h^2(\cdot)\in L_1(E_1)$ we can form the bi-distributive functional

$$P(h^1,h^2)\equiv E\left\{\int_{-\infty}^{\infty}\overline{h^1(r)}\,x(r)\,dr\,.\int_{-\infty}^{\infty}h^2(s)\,\overline{x(s)}\,ds\right\} \tag{6.6.4}$$

$$\equiv\int_{-\infty}^{\infty}\int_{-\infty}^{\infty}\overline{h^1(r)}\,h^2(s)\,R(r-s)\,dr\,ds, \tag{6.6.5}$$

and if we introduce the transforms

$$g^m(\alpha)=\int_{-\infty}^{\infty}e^{-2\pi it\alpha}\,h^m(t)\,dt,\quad m=1,2, \tag{6.6.6}$$

then we have

$$P(h^1,h^2)=\int_{-\infty}^{\infty}\int_{-\infty}^{\infty}\overline{g^1(\alpha)}\,g^2(\alpha)\,d\Gamma(\alpha). \tag{6.6.7}$$

Proof. $\|x(t)\|^2=E\{x(t)\overline{x(t)}\}=R(0)$ implies in particular that $\|x(t)\|$ is bounded, and since $x(t)$ is also continuous in norm, we can consider the integral

$$\int_{-\infty}^{\infty}\overline{h(t)}\,x(t)\,dt \tag{6.6.8}$$

for any $h(\cdot)\in L_1(E_1)$, and it is even possible to define the integral

$$\int_{-\infty}^{\infty}x(t)\,\overline{dH(t)} \tag{6.6.9}$$

as a Riemann integral for any $H(\cdot)\in V(E_1)$. We can also form (6.6.4) and since for $h\in L_1$ we have

$$0\leqq P(h,h)=\int_{-\infty}^{\infty}\int_{-\infty}^{\infty}\overline{h(r)}\,h(s)\,R(r-s)\,dr\,ds,$$

we can indeed apply our theorem 3.2.3, and a representation (6.6.2) with (6.6.3) exists indeed. Also we can substitute (6.6.2) in (6.6.5) and (6.6.7) ensues, as claimed.

Next, (6.6.8) is a distributive functional from L_1 to $L_2(\Omega)$, and if we introduce the transforms

$$g(\alpha)=\int_{-\infty}^{\infty}h(t)\,e^{-2\pi it\alpha}\,dt, \tag{6.6.10}$$

we may also view it as a distributive functional from the vector space $\{g(\cdot)\}$ to $L_2(\Omega)$. Now, formally such a functional can be represented as an integral $\int_{-\infty}^{\infty} \overline{g(\alpha)}\, dY(\alpha)$ with $Y(\Delta) \in L_2(\Omega)$, so that we have

$$\int_{-\infty}^{\infty} \overline{h(t)}\, x(t)\, dt = \int_{-\infty}^{\infty} \overline{g(\alpha)}\, dY(\alpha), \tag{6.6.11}$$

and if we denote the transforms of h^1, h^2 by g^1, g^2, then (6.6.4) is formally

$$Q(g^1, g^2) = E\left\{ \int_{-\infty}^{\infty} \overline{g^1(\alpha)}\, dY(\alpha) \cdot \int_{-\infty}^{\infty} g^2(\beta)\, d\overline{Y}(\beta) \right\}, \tag{6.6.12}$$

just as in section 6·5. But formally this implies the orthogonality relation

$$E\{Y(\Delta_1)\, \overline{Y(\Delta_2)}\} = \Gamma(\Delta_1 \cap \Delta_2), \tag{6.6.13}$$

and the fact is that all this can be made rigorous and that the following continuation to the preceding theorem can be stated:

THEOREM 6.6.2. *Also, there exists, on all bounded intervals, a finitely additive function* $Y(\Delta) \in L_2(\Omega)$ *for which* (6.6.13) *and* (6.6.11) *holds.*

Conversely, if we are given a function $Y(\Delta)$ *for which* (6.6.13) *with* (6.6.3) *holds then there exists a function* $x(t)$ *as in theorem* 6.6.1 *for which* (6.6.11) *holds.*

We will not reproduce the proof.

Remark 1. Relation (6.6.11) can be enlarged to the more self-dual one:

$$\int_{-\infty}^{\infty} x(t)\, \overline{dH(t)} = \int_{-\infty}^{\infty} \overline{g(\alpha)}\, dY(\alpha), \tag{6.6.14}$$

where $H(\cdot) \in V(E_1)$, and $g(\alpha)$ is its transform.

Remark 2. If a σ-additive $\Gamma(A)$ is finite on bounded sets, and if there is a $Y(\Delta)$ with property (6.6.13), and if we introduce the decomposition

$$\Gamma(A) = \Gamma_1(A) + \Gamma_2(A) + \Gamma_3(A)$$

into discrete, singular-continuous and absolutely continuous addends, then $Y(\Delta)$ can be written as $Y_1(\Delta) + Y_2(\Delta) + Y_3(\Delta)$, where $Y_m(\Delta)$ satisfies (6.6.13) for Γ_m, $m = 1, 2, 3$. Hence the following further statement:

THEOREM 6.6.3. *Finally, we can put uniquely* $x(t) = x_1(t) + x_2(t) + x_3(t)$, *where each* $x_m(t)$ *is K-stationary and has a transform* $Y_m(\Delta)$ *as in remark* 2 *of section* 6.6.

Any σ-additive $\Gamma(A) \geqq 0$ with

$$\Gamma(E_1) < \infty \tag{6.6.15}$$

can be the covariance function of a function $Y(\Delta)$, compare theorem 6.1.2, and thus, except for the restriction (6.6.15), our present theorem is much more comprehensive than was theorem 6.5.1 in which $\Gamma(\Delta)$ had to be absolutely continuous always. This being so, we will further note that even the restriction (6.6.15) can be removed, and that there is a proposition available, which can be stated in several versions, of which both of our previous theorems are special cases.

THEOREM 6.6.4. *If* $\{x^\epsilon(t)\}$ *in* $0<\epsilon<\infty$ *is a family of stationary random point functions as in theorems* (6.6.1)–(6.6.3), *and if for any* $\epsilon>0$, $\eta>0$ *we have*

$$x^{\eta+\epsilon}(t)=\int_{-\infty}^{\infty}\frac{1}{\sqrt{\eta}}\exp\left[-\frac{\pi}{\eta}(t-\tau)^2\right]x^\epsilon(\tau)\,d\tau, \qquad (6.6.16)$$

then there exists a representation

$$x^\epsilon(t)\sim\int_{-\infty}^{\infty}e^{2\pi it\tau}\,e^{-\epsilon\pi\tau^2}dY(\tau) \qquad (6.6.17)$$

by a joint orthogonal function $Y(\Delta)$, *meaning that if we introduce the representations*

$$x^\epsilon(t)\sim\int_{-\infty}^{\infty}e^{2\pi it\tau}\,dY^\epsilon(\tau) \qquad (6.6.18)$$

of theorem 6.6.2 *then we have*

$$Y^\epsilon(\Delta_{\alpha\beta})=\int_\alpha^\beta e^{-\epsilon\pi\tau^2}dY(\tau). \qquad (6.6.19)$$

Proof. If we apply (6.6.11) with $x(r)\equiv x^\epsilon(r)$, $Y(\Delta)\equiv Y^\epsilon(\Delta)$ and $h(r)=\frac{1}{\sqrt{\eta}}\exp\left[-\frac{\pi}{\eta}(t-r)^2\right]$ then we obtain

$$\int_{-\infty}^{\infty}\frac{1}{\sqrt{\eta}}\exp\left[-\frac{\pi}{\eta}(t-r)^2\right]x^\epsilon(r)\,dr=\int_{-\infty}^{\infty}e^{2\pi it\tau-\pi\eta\tau^2}dY^\epsilon(\tau),$$

but since by explicit assumption the left side is

$$f^{\eta+\epsilon}(t)\sim\int_{-\infty}^{\infty}e^{2\pi it\tau}\,dY^{\eta+\epsilon}(\tau),$$

we therefore obtain $Y^{\eta+\epsilon}(\Delta_\alpha)=\int_\alpha e^{-\pi\eta\tau^2}dY^\epsilon(\tau)$

for any $\eta>0$, $\epsilon>0$. From this it is possible to conclude that we have

$$\int_\alpha^\beta e^{\pi(\eta+\epsilon)\tau^2}dY^{\eta+\epsilon}(\tau)=\int_\alpha^\beta e^{\pi\epsilon\tau^2}dY^\epsilon(\tau),$$

for all $\eta > 0$, $\epsilon > 0$, and this means that the integral

$$Y(\Delta) = \int_\alpha^\beta e^{\pi\epsilon\tau^2} dY^\epsilon(\tau)$$

is independent of ϵ, so that (6.6.19) holds, which proves the theorem.

Now, if we take a function $x(t)$ to which theorems (6.6.1)–(6.6.3) apply, and if we form

$$x^\epsilon(t) = \int_{-\infty}^{\infty} \frac{1}{\sqrt{\epsilon}} \exp\left[-\frac{\pi}{\epsilon}(t-\tau)^2\right] x(\tau)\, d\tau,$$

then this is a family to which theorem 6.6.4 applies, with $Y(\Delta)$ being the same as in theorem 6.6.2. Also $x(t)$ is a limit in norm of $x^\epsilon(t)$ as $\epsilon \downarrow 0$.

On the other hand, if we take a set function $X(\Delta)$ as in theorem 6.5.1 and put

$$x^\epsilon(t) = \int_{-\infty}^{\infty} \frac{1}{\sqrt{\epsilon}} \exp\left[-\frac{\pi}{\epsilon}(t-\tau)^2\right] dX(\tau),$$

then this is also a family to which theorem 6.6.4 applies with $Y(\Delta)$ being the same as in theorem 6.5.1. Furthermore, it is easy to show from $L_{2,2}$-boundedness that

$$X^\epsilon(\Delta_{\alpha\beta}) \equiv \int_\alpha^\beta x^\epsilon(t)\, dt$$

converges in norm to $X(\Delta_{\alpha\beta})$ as $\epsilon \downarrow 0$, and this theorem 6.6.4 does indeed include both previous ones, in a sense.

Note that the function $x^\epsilon(t)$ is a random-valued solution of the equation

$$\frac{\partial^2 x^\epsilon(t)}{\partial t^2} = 4\frac{\partial x^\epsilon(t)}{\partial \epsilon},$$

and that we could have also taken the 'harmonic' equation instead, in which case we would have had

$$x^\epsilon(t) \sim \int_{-\infty}^{\infty} e^{2\pi i t\tau - 2\epsilon\pi|\tau|} dY(\tau),$$

but we will not pursue this topic any further.

NOTES AND REFERENCES

Chapter 1

1.1. General approximation theorems are also to be found in E. W. Hobson, *Theory of Functions of a Real Variable*, vol. 2 (Cambridge University Press, 1947).

1.2. Translation functions, after having been mentioned incidentally by H. Bohr, were introduced systematically in the second half of our paper, 'Beiträge zur Theorie der fast periodischen Funktionen. I', *Math. Ann.* vol. 96 (1926), pp. 119–47.

For Stepanoff functions see, for instance, A. S. Besicovitch, *Almost Periodic Functions* (Cambridge University Press, 1932).

1.3. The Hölder-Minkowski inequality here used is formula (6.13.9) on p. 148 in G. H. Hardy, J. E. Littlewood and J. Pólya, *Inequalities* (Cambridge University Press, 1934).

1.4. The Lebesgue integral of Banach-valued functions was introduced in our paper, 'Integration von Funktionen, deren Werte die Elemente eines Vektorraumes sind', *Fund. Math.* vol. 20 (1933), pp. 262–76. See also T. H. Hildebrandt, 'Integration in abstract spaces', *Bull. Amer. Math. Soc.* vol. 59 (1953), pp. 111–39, and pp. 40–52 in E. Hille, *Functional Analysis and Semi-groups* (Amer. Math. Soc., Providence, R.I., 1948). For additional Fourier analytic applications of the integral see S. Bochner and A. E. Taylor, 'Linear functionals on certain spaces of abstractly valued functions', *Ann. Math.* vol. 39 (1938), pp. 262–76, and R. P. Boas and S. Bochner, 'On a theorem of M. Riesz for Fourier series', *J. Lond. Math. Soc.* vol. 14 (1939), pp. 62–73.

1.5. For additive set functions in Euclidean space see E. J. McShane *Integration* (Princeton University Press, 1944); J. Radon, 'Theorie und Anwendungen der absolut additiven Mengen funktionen', *S.B. Akad. Wiss. Wien*, vol. 122 (1913), pp. 1293–439; Hans Hahn, *Reelle Funktionen, I* (Springer, Berlin, 1921); and S. Bochner, 'Monotone Funktionen, Stieltjessche Integrale und harmonische Analyse', *Math. Ann.* vol. 10 (1933), pp. 378–410.

Theorem 1.5.4 is due to A. Plessner, 'Eine Kennzeichnung der totalstetigen Funktionen', *J. Reine Angew. Math.* vol. 160 (1929), pp. 26–32. The extension to Haar measure on compact groups was given in our paper, 'Additive set functions on groups', *Ann. Math.* vol. 40 (1939), pp. 769–96 (theorem 15 on p. 789).

Chapter 2

2.2. When a course of lectures on this topic was being given in the spring of 1953 at the Statistical Laboratory of the University of California, Berkeley, a remark made by Dr E. Parzen led to theorem 2.2.2 becoming as general as it is now, and a remark made by Dr H. Flanders led to theorem 4.1.3 becoming as general as it is now.

For a discussion and proof of theorem 2.2.4 see S. Bochner and K. Chandrasekharan, *Fourier Transforms* (Princeton University Press, 1949: *Ann. Math. Studies*, no. 19).

2.4. There are possibilities of extending the Poisson summation formula from sums over strict lattices to sums over more general point sets. For one variable some such statements were made in our paper, 'A generalization of Poisson's summation formula', *Duke Math. J.* vol. 6 (1940), pp. 229–34, and for several variables the problem was mentioned in 'On spherical partial sums of multiple Fourier series', *Rev. Cienc., Lima* (1948), pp. 85–104.

Formulas for Fourier integrals used in the text will be found in *Vorlesungen über Fouriersche Integrale* (Chelsea, New York, 1948).

2.5. For 'spherical summability' see our paper, 'Summation of multiple Fourier series by spherical means', *Trans. Amer. Math. Soc.* vol. 40 (1936), pp. 175–207 and also chapter IV in the book by K. Chandrasekharan and S. Minakshisundaram, *Typical Means* (Oxford University Press, 1952: Tate Institute of Fundamental Research, Monographs on Mathematics and Physics).

The diffusion equation with general completely monotone transformations of the Laplacean which we will yet discuss in section 4.6, especially with fractional powers of the Laplacean and the link to symmetric stable processes, and also the subordination of Markoff processes which we will discuss in section 4.4, were introduced in our papers, 'Quasi-analytic functions, Laplace operator, positive kernels', *Ann. Math.* vol. 51 (1950), pp. 68–91 [this paper contains details and applications that will not be mentioned in the text], and 'Diffusion equation and stochastic processes', *Proc. Nat. Acad. Sci., Wash.*, vol. 35 (1949), pp. 368–70. The general transformations of the Laplacean by themselves were already introduced in 'Completely monotone functions of the Laplace operator for torus and sphere', *Duke Math. J.* vol. 3 (1937), pp. 488–502.

The case of fractional powers of the Laplacean in $-\infty < x < \infty$ was linked to Riemann–Liouville integrals in W. Feller, 'On a generalization of Marcel Riesz's potentials and the semi-groups generated by them', *Comm. Sém. Math. Univ. Lund*, Tome supplémentaire (1952), pp. 73–81.

2.6–2.8. These sections reproduce, with additions, the major part but not all of the contents of the following two papers of ours: 'Theta relations with spherical harmonics', *Proc. Nat. Acad. Sci., Wash.*, vol. 37 (1951), pp. 804–8; 'Zeta functions and Green's functions for linear partial differential operators of elliptic type with constant coefficients', *Ann. Math.* vol. 57 (1953), pp. 32–56.

Chapter 3

3.1–3.4. These sections are an elaboration of our note, 'Closure classes originating in the theory of probability', *Proc. Nat. Acad. Sci., Wash.*, vol. 39 (1953), pp. 1082–8.

In $-\infty < x < \infty$, the famed structure theorem 3.4.2 was conceived, with an imperfection, in A. Kolmogoroff, 'Sulla forma generale di un processo stocastico omogeneo', *R.C. Accad. Lincei*, vol. 15 (6), 1932, pp. 805–8, 866–9. The imperfection was soon removed by Paul Lévy, and the theorem was variously reproven, and a systematic analysis of this (one-dimensional) theorem will be found in M. Loève, 'On sets of probability laws and their limit elements', *Univ. California Publ. Statist.* vol. 1, no. 5 (1952), pp. 53–88. This paper is also, in a sense, concerned with the peculiar relationship of this theorem to the central limit theorem, and on this some brief comment had already been made in W. Feller, 'The fundamental limit theorems in probability', *Bull. Amer. Math. Soc.* vol. 51 (1945), pp. 800–32.

The structure theorem 3.4.2 for several space variables was set up in P. Lévy, *Théorie de l'addition des variables aléatoires* (Gautheir-Villars, Paris, 1937), pp. 212–20.

3.5. For an extension of the theorems to several variables see our paper, 'Stochastic processes with finite and nonfinite variance', *Proc. Nat. Acad. Sci., Wash.*, vol. 39 (1953), pp. 190–7.

For functions of slow growth see J. Karamata, 'Sur un mode de croissance regulière des fonctions', *Mathematica, Cluj*, vol. 4 (1930), pp. 38–53.

3.8. See S. Bochner, 'Completely monotone functions in partially ordered spaces', *Duke Math. J.* vol. 9 (1942), pp. 519–26; S. Bochner and Ky Fan, 'Distributive order preserving operations in partially ordered vector spaces', *Ann. Math.* vol. 48 (1947), pp. 168–9, and the last chapter of the *Ann. Math. Studies* by E. J. McShane, *Order Preserving Maps and Integration Processes* (Princeton University Press, 1952).

For the result of H. Cramér see his paper 'On the theory of stationary random processes', *Ann. Math.* vol. 41 (1940), pp. 215–30.

Chapter 4

4.1. One half of theorem 4.1.2 was given in our paper 'Stable laws of probability and completely monotone functions', *Duke Math. J.* vol. 3 (1937), pp. 726–8, and we used it in effect to show that the symmetric stable distributions are subordinate to the Gaussian in the sense of section 4.3. The second half of theorem 4.1.2 was then stated, for another purpose, in I. J. Schoenberg, 'Metric spaces and completely monotone functions', *Ann. Math.* vol. 39 (1938), pp. 811–41.

4.2. Theorem 4.2.2 was obtained in a Princeton doctoral thesis by W. Gilbert, 1952.

4.4. The notion of continuity as introduced in definition 4.4.2 is perhaps ill named because it is quite different from the one introduced in A. Kolmogoroff, 'Über die analytischen Methoden in der Wahrscheinlichkeitsrechnung', *Math. Ann.* vol. 104 (1931), pp. 415–58, and in W. Feller, 'Zur theorie der stochastischen Prozesse', *Math. Ann.* vol. 113 (1936), pp. 113–60. This latter continuity means more or less that almost all paths are continuous, and this in turn means more or less that in the associated diffusion equation the operator in the space variables is a partial differential operator of second order and in the case of a stationary process even a 'genuine' Laplacean.

We might mention that an important problem on diffusion was studied for fractional powers of the Laplacean in M. Kac, 'On some connections between probability theory and differential and integral equations', *Proceedings of the Second Berkeley Symposium on Mathematical Statistics and Probability* (University of California Press, 1951, pp. 189–215.)

To theorem 4.4.4 compare T. J. Schoenberg, 'Metric spaces and positive definite functions', *Trans. Amer. Math. Soc.* vol. 44 (1938), pp. 522–36.

4.5. See R. E. A. C. Paley, 'A proof of a theorem on averages', *Proc. Lond. Math. Soc.* (2), vol. 31 (1930), pp. 289–300.

4.8 and **4.9.** These sections summarize part of the results from the following papers of ours: 'Some properties of modular relations', *Ann. Math.* vol. 53 (1951), pp. 332–63, 'Connection between functional equations and modular relations, and functions of exponential type', *J. Indian Math. Soc.* vol. 16 (1952), pp. 99–102; 'Bessel functions and modular relations of higher type and hyperbolic differential equations', *Comm. Sém. Math. Univ. Lund*, Tome supplémentaire (1952), pp. 12–20.

Chapter 5

The content of this and the next chapter is meant either to supplant or elaborate most, though not all, of the results in the following interlocking papers of ours: 'Stochastic processes', *Ann. Math.* vol. 48 (1947), pp. 1014–61;

'Partial ordering in the theory of stochastic processes', *Proc. Nat. Acad. Sci., Wash.*, vol. 36 (1950), pp. 439–43; 'Length of random paths in general homogeneous spaces', *Ann. Math.* vol. 57 (1953), pp. 309–13; 'Fourier transforms of time series', *Proc. Nat. Acad. Sci., Wash.*, vol. 39 (1953), pp. 302–7.

5.1. For directed sets and inverse mapping systems see, for instance, S. Lefschetz, *Algebraic Topology* (Amer. Math. Soc., Providence, R.I., 1942), pp. 31–3 and other pertinent passages.

In connection with theorem 5.1.2, the author is indebted for enlightening advice to Dr Lucien LeCam and also to Dr L. Breiman.

5.4. The characteristic functional was systematically introduced in our 'Stochastic Processes' *Ann. Math.* vol. 48 (1947), pp. 1014–61, for random additive set functions, and it was also introduced for point functions in L. LeCam, 'Un instrument d'étude des fonctions aléatoires: la fonctionelle caractéristique', *C.R. Acad. Sci., Paris*, vol. 224 (1947), pp. 710–11. For applications see D. G. Kendall, 'Stochastic processes and population growth', *J.R. Statist. Soc.* B, vol. 11 (1949), pp. 230–65; M. S. Bartlett and D. G. Kendall, 'On the use of the characteristic functional in the analysis of some stochastic process occurring in physics and biology', *Proc. Camb. Phil. Soc.* vol. 47 (1951), pp. 65–80; and also M. S. Bartlett, 'The dual recurrence relation for multiplicative processes', *Proc. Camb. Phil. Soc.* vol. 47 (1951), pp. 821–5.

5.5. Dealing virtually exclusively with random functions on the ordinary straight line, J. L. Doob, both in his papers and in his treatise, *Stochastic Processes* (Wiley, New York, 1945), prefers to speak of a 'process of increments' rather than of random interval functions.

5.6. See G. E. Bates and Jerzy Neyman, 'Contributions to the theory of accident proneness. II. True or false contagion', *Univ. California Publ. Statist.* vol. 1 (1952), pp. 255–76, in particular formula (56) and neighboring ones. This paper was enlightening to us for the emphasis it places on the fact that in population and fission problems the random entity is not a point function but a finitely additive interval function only, and that care must be taken in distinguishing which end point of the interval does belong to it and which does not.

Chapter 6

6.1. For random-point functions on arbitrary point set, theorems 6.1.3 and 6.1.2 were presented in the memoir of M. Loève, 'Fonctions aléatoires du second order', which was published on pp. 299–352 as a 'Note' to the book by P. Lévy, *Processus Stochastiques et Mouvement Brownien* (Gauthier-Villars, Paris, 1949).

Theorem 6.1.2, again for point functions only and with further restriction of 'separability', was also implicitly proven by Hilbert space theory on pp. 368–71 of N. Aronszajn, 'Theory of reproducing kernels', *Trans. Amer. Math. Soc.* vol. 68 (1950), pp. 337–404.

6.2. What we now call L-finiteness was in our 'Stochastic Processes', *Ann. Math.* vol 48 (1947), pp. 1014–61, called L-stability.

6.2–6.5. The spectral theory of a stationary process was begun in A. Kintchine, 'Korrelationstheorie der stationären stochastischen Prozesse', *Math. Ann.* vol. 190 (1934), pp. 604–15, and a rounded version of Kintchine's theorem was given in H. Cramér, 'On harmonic analysis in certain functional spaces', *Ark. Mat. Astr. Fys.* vol. 28B, no. 12 (1942), 17 pp., but in the meantime H. Wold, in *A Study in the Analysis of Stationary Time Series* (Almquist and Wiksells, Uppsala, 1938), had taken the first systematic steps towards utilizing the spectral resolution for purposes of 'prediction theory'.

After 1945 the spectral theory was reexamined in several studies, virtually simultaneously and independently of one another. Loève introduced, among others, the (nonstationary) 'harmonizable' process. K. Karhunen in 'Über lineare Methoden in der Wahrscheinlichkeitsrechnung', *Ann. Acad. Sci. Fenn.* A, vol. 37 (1947), 79 pp., very much exploited the Hilbert space approach. J. L. Doob in his article, 'Time series and harmonic analysis', *Proceedings of the Berkeley Symposium in Mathematical Statistics and Probability* (University of California Press, 1949, pp. 303–43), stressed the viewpoint of Norbert Wiener and finally in our 'Stochastic processes' *Ann. Math.* vol. 48 (1947), pp. 1014–61, the first attempt was made (still abortively there) towards dualizing the theory by envisaging interval functions on both sides of the inversion formula.

GENERAL INDEX

INDEX OF SYMBOLS